SECOND EDITION

Residential Concrete

by the NAHB Research Foundation, Inc.,
revised and updated by Harold W. Conner

Home Builder Press®
National Association of Home Builders
1201 15th Street, NW
Washington, DC 20005-2800

Residential Concrete, second edition

ISBN 0-86718-389-6

The following illustrations and photographs are reprinted with permission from the persons or organizations noted: Plot plan from W.P. Jackson, *Building Layout*, 1990, courtesy of Craftsman Book Co.; builder's level, courtesy of Berger instruments; power finisher-float illustration from the *Concrete Craftsman Series: Slabs on Grade*, courtesy of the American Concrete Institute; onsite photograph of reusable plywood and steel panel forming system, courtesy of Dave Parrot, Orvin Construction; steel-ply panels and fillers, courtesy of the Symons company; table on "Maximum Spacing of Control Joints," courtesy of Portland Cement Association; weathering map and associated text from draft material for American Concrete Institute Committee 332's updated *Guide to Residential Cast-in-Place Concrete*, courtesy of ACI; concrete tools photograph, courtesy of the Portland Cement Association; and the many photographs illustrating common concrete problems and creative finishes and styles, courtesy of the Portland Cement Association.

Printed in the United States of America

Library of Congress Cataloging-in-Publication Data

Residential concrete / by the NAHB Research Foundation, Inc. — 2nd ed. / rev. and updated by Harold W. Conner.
p. cm.
Rev. ed. of: Residential concrete / Laurence Miller. 1983.
ISBN 0-86718-389-6
1. Concrete construction. I. Conner, Harold W. II. NAHB Research Foundation. III. Miller, Laurence. Residential concrete.
TA682.4.R47 1994
691'.3—dc20 93-36745
CIP

For further information, please contact—

Home Builder Press®
National Association of Home Builders
1201 15th Street, NW
Washington, DC 20005-2800
(800) 223-2665

12/93 EPS/United 2,500
10/94 United reprint 3,000
7/97 United reprint 1,000

Contents

Preface

This second edition of *Residential Concrete* covers virtually everything the home builder needs to know about high-quality concreting. While not a training manual, the book describes basic techniques and provides guidelines for ordering ready mixed concrete, working with admixtures (such as accelerators, retarders, water reducers, and high-range water reducers), forming, jointing, curing, and controlling basement leakage. Many sections include detailed illustrations.

Curing is given special attention because poor curing is the source of many common concrete problems. Proper curing greatly strengthens concrete and increases its watertightness. All the basic curing methods are discussed, including their advantages and disadvantages.

A comprehensive section is devoted to common concrete problems and their remedies. In fact, the problems of scaling and spalling were the primary reasons the National Association of Home Builders' Standing Committee on Research commissioned the original edition of this publication in the early 1980s. Plastic-shrinkage cracks, blisters, and popouts are among the problems covered. Patching techniques using latex, epoxy, and dry-mix concrete are provided. Throughout the book, special measures for cold weather and for hot, dry, windy weather are discussed. The detailed list of contents enables the reader to turn directly to any topic covered.

In revising and updating this second edition, special attention has been paid to new developments in forms, changes in concrete admixtures, and up-front concerns such as planning the site layout and designing with due consideration to soil conditions. The chapters have been subdivided and reorganized to make the book easier to use as a reference. A list of sources for additional information and a glossary of basic concrete terminology have been added.

Acknowledgments

The first edition of *Residential Concrete* was written and illustrated by Laurence Miller of the NAHB Research Foundation, Inc. under the guidance of Lee Fisher and Ralph J. Johnson. Reviewers of the first edition included the Technical Services Department of the National Association of Home Builders and builder members of the NAHB Standing Committee on Research.

The author of the revisions and updates in this second edition of *Residential Concrete* is Harold W. Conner, Associate Professor and Director of Construction Science at the University of Oklahoma, Norman. Professor Conner's background includes experience as a contractor, concrete specialist, and consultant to builders, and publication of numerous articles on concrete topics in industry publications.

Content reviews for the second edition were kindly provided by the following individuals: Pat Persico and staff engineers at Master Builders, Inc., Cleveland, Ohio; William Panarese, Jamie Farny, and Dan Mistick, Portland Cement Association, Skokie, Ill.; James Carr, Construction Technologies Coordinator, The Ohio State University, Agricultural Tech. Institute, Wooster, Ohio; Donald Luebs, NAHB Research Center, Upper Marlboro, Md.; David Parrot, Orvin Construction, Bremerton, Wash.; and the following members of the National Ready Mixed Concrete Association's Residential Concrete Task Group—Thomas Adams, Michigan Concrete Association, Lansing, Mich.; Larry J. Asel, Concrete Company of Springfield, Springfield, Mo.; Nicholas Maloof, Thomas Concrete of Georgia, Inc., Atlanta, Ga.; Dale Rech, Owl Rock Products, Arcadia, Calif.; Andrew W. Young, Blue Circle, Macon, Ga.; Robert A. Garbini, NRMCA, Silver Spring, Md.; and Tarek S. Khan, NRMCA, Silver Spring, Md. Thanks also go to Marty Taylor, who copy-edited this second edition.

This book was produced under the general direction of Kent Colton, NAHB Executive Vice President, in association with NAHB staff members James DeLizia, Staff Vice President, Member and Association Relations Division; Adrienne Ash, Assistant Staff Vice President, Publishing Services; Rosanne O'Connor, Director of Publications; Sharon Lamberton, Assistant Director of Publications and Project Editor; David Rhodes, Art Director; Julie Wilson, Marketing Director; and Carolyn Poindexter, Editorial Assistant.

Chapter 1

Understanding Concrete

Concrete Basics

Cement Hydration and Water

Cement particles, through a chemical process called hydration, develop long crystals when wet, and crystallization bonds the concrete mix together.

As long as moisture is present and the temperature is not too low, the crystals can continue to grow for years and increase the strength of the concrete. Cement needs only a small amount of water to hydrate. Too much water will seriously weaken the concrete.

The water-to-cement ratio is important. Typically, the ratio should be 0.50 or less by weight—or no more than ½ pound of water to 1 pound of cement. Moisture in the sand must always be included when measuring the water. The maximum 1 to 2 ratio is necessary for durability, watertightness, and strength. If the concrete is too stiff and difficult to work at this ratio, a water-reducing admixture may be used.

Problems from Too Much Water

Too much water in the mix can create water reservoirs in the concrete and cause it to bleed excessively. (Bleeding occurs when excess water rises to and

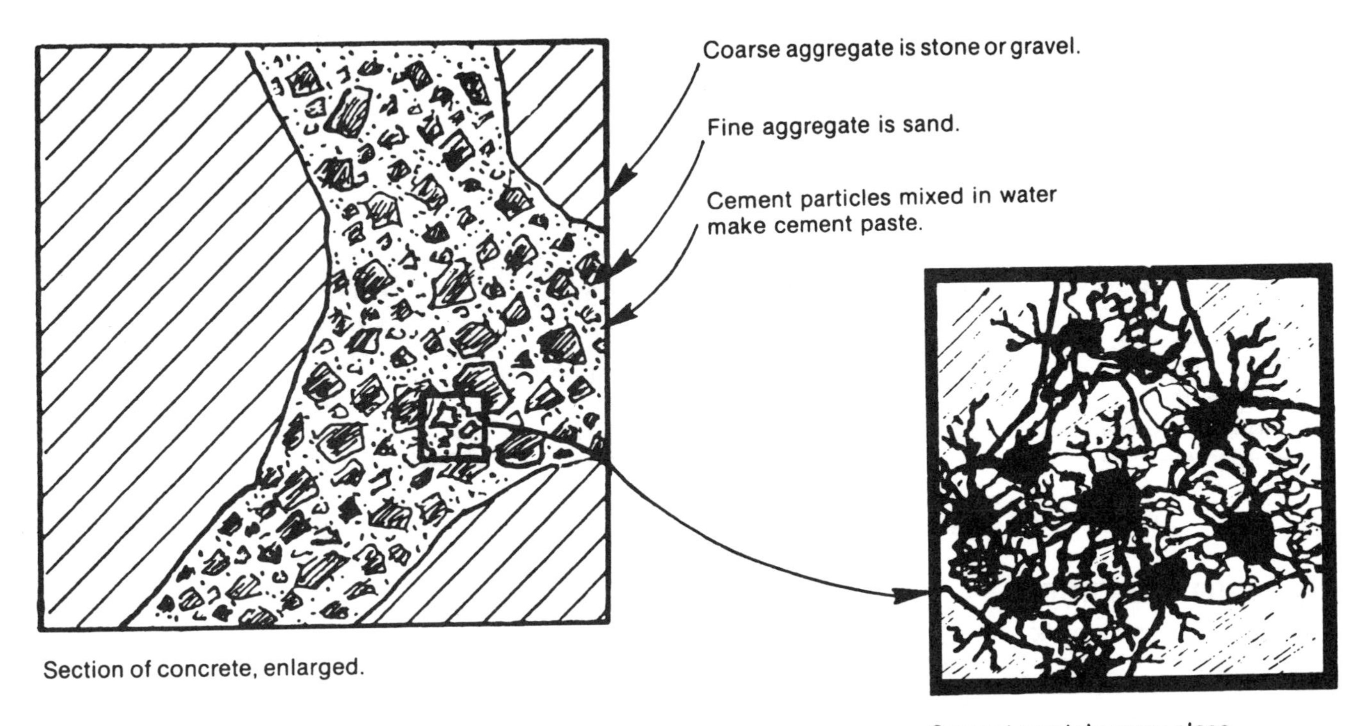

Section of concrete, enlarged.

Cement crystals grow close together in properly mixed concrete.

When too much water is used, the cement crystals are far apart and the concrete is much weaker.

A water-to-cement ratio that is too high cuts concrete strength and greatly affects watertightness.

accumulates on the top surface of the concrete.) Channels lead from the reservoirs to the surface where a watery laitance forms and a weak surface results. The concrete may craze, dust, and scale, especially if finishing operations work the bleed water back into the surface before it can evaporate. If the water reservoirs and channels dry out, air pockets will remain and allow water and deicers to enter. Should the water freeze and expand, it will severely test the durability of the concrete. Durability is resistance of the concrete to freeze-thaw cycles, chemical attack, abrasion, and other conditions of service. Deicers also may seriously harm concrete that is not air entrained.

Air entrainment lessens bleeding and helps protect the concrete against freeze-thaw damage and deicers. However, even entrained air loses some of its effectiveness in a soupy mix.

Problems from Too Little Water

It is almost impossible to make concrete with too little water, since the minimum required for the chemical hydration of cement would not produce a workable mix. However, water that is added to the mix should be thoroughly blended into the entire batch of concrete. Inadequate mixing can cause serious segregation of the ingredients, compromising the quality of the finished concrete.

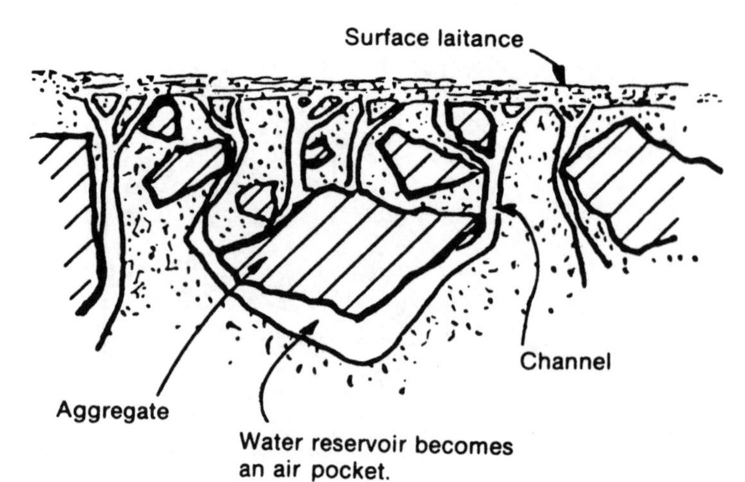

Hydration and Curing

As long as concrete is allowed to cure, it continues to gain strength and watertightness for years after it is placed. The greatest strength gain, however, occurs in the first week or two after placing—and it is then that curing is most important.

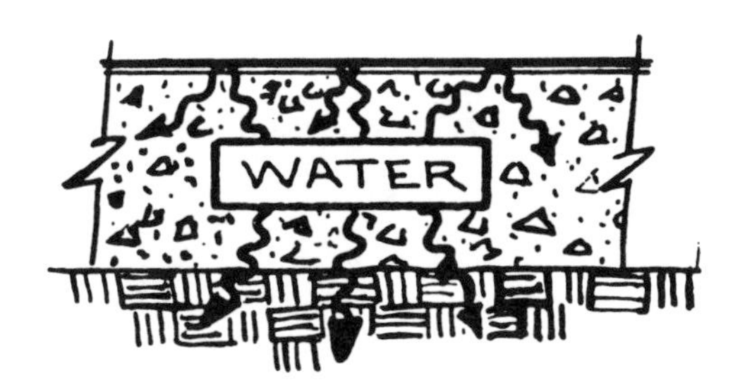

Subgrade should be properly moistened before placing concrete so that too much water from the mix does not go into the subgrade. A good cure will retain water in the concrete.

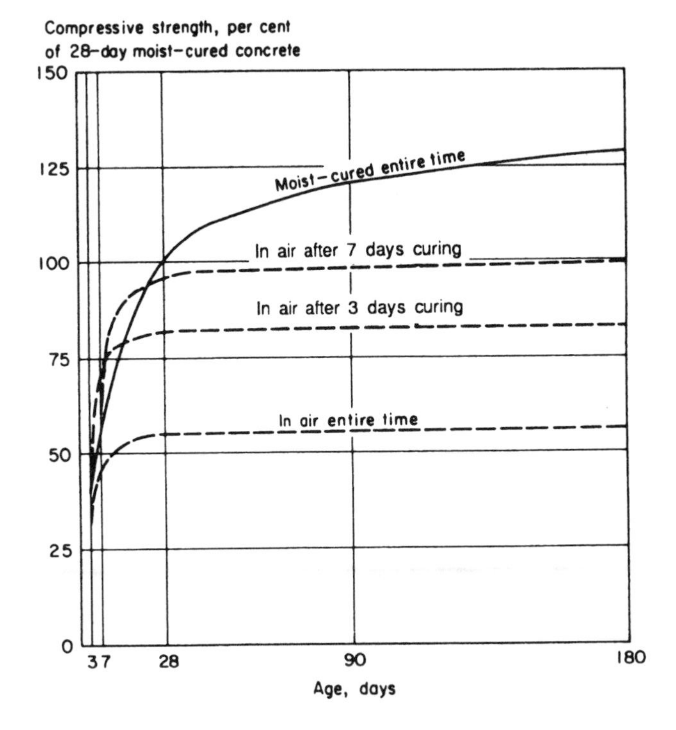

Concrete strength increases with age as long as moisture and favorable temperatures are present. Good curing more than doubles the strength of concrete.

Uncured concrete dries out and may reach only half its design strength and show reduced durability. Proper curing requires that the concrete be kept moist and at a moderate temperature. Curing methods are discussed in more detail in Chapter 5.

In addition to increasing the strength of the concrete, curing also reduces its shrinkage stress. A combination of increased strength and stress reduction means fewer cracks.

> **Basic Recipe for Quality Concrete**
> A low water-cement ratio and good curing combined will yield a stronger finished concrete with increased watertightness and better resistance to spalling and cracking.

Mix Design

The right proportions of all ingredients, not just the water-to-cement ratio, must be maintained. For example—

- Too much cement paste is not only costly, but will produce concrete that is less resistant to abrasion and cracks more easily.
- Too much aggregate (sand and stone) makes the mix stiff or harsh and difficult to place and finish. Too much fine aggregate (sand) requires excess water, which may lead to cracking and excessive shrinkage. Too much coarse aggregate (stone) produces a porous and honeycombed concrete.

For guidelines on specifying ready mixed concrete, see Chapter 4.

Slump

The slump test, illustrated here, is a measure of the consistency, stiffness, and workability of fresh concrete. It is influenced by the amount of water—more water means higher slump—but water is not the only influence. The type of aggregate, the air content, the admixtures, temperature, and the proportions of all the ingredients affect slump. Mixing time and standing time also affect slump.

Typical slumps for various jobs are given in the mix design table on page 36.

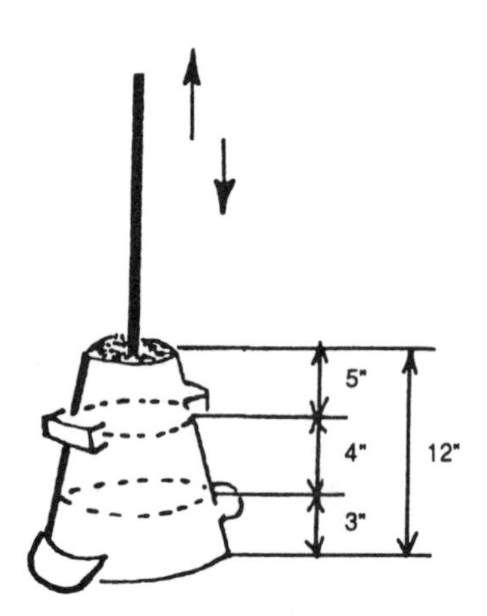

Slump test. Step 1. Fill the cone in three layers of equal volume, rodding each layer 25 times.

Step 2. Strike off the top, then remove the cone slowly with an even motion, taking from 5 to 12 seconds. Do not jar the mixture or tilt the cone in the process.

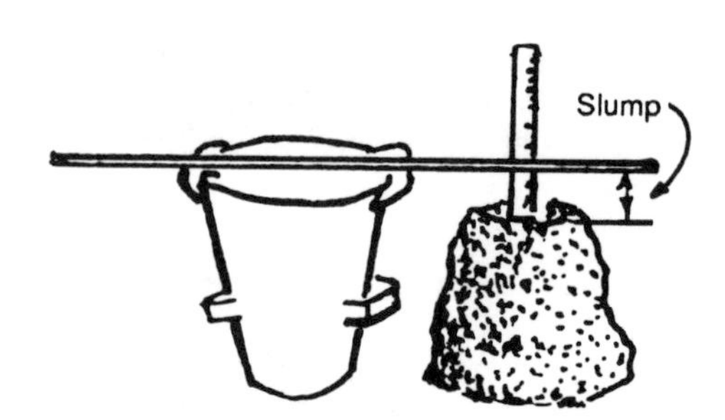

Step 3. Measure the slump with tamping rod and ruler. The slump test should not take longer than 1½ minutes. Do not use the same batch of concrete for any other test.

Admixtures

All admixtures (ingredients added before or during mixing) used should meet American Society for Testing and Materials (ASTM) Standards Designation C 494. Admixtures are used to strengthen concrete, to speed up or slow down the set-up time, and to help protect concrete against the effects of temperature changes and exposure to chemicals such as deicers.

Air Entrainment

Air entrainment helps protect concrete that will be exposed to freezing and thawing and deicers.

An air entrainment admixture causes microscopic air bubbles to form throughout the concrete. These tiny bubbles function as relief valves when water in the concrete freezes, helping to prevent scaling or spalling of the surface. Resistance to deicers, which also cause scaling and spalling, is greater if the air-entrained concrete is air-dried for about four weeks after curing.

Air-entrained concrete is more watertight, more resistant to sulfate soils, and easier to work—provided the right tools are used—particularly if the mixture is lean or has angular aggregates. Concrete strength is reduced somewhat by air entrainment, but a lower water-cement ratio is possible in an air-entrained mix and this usually balances lost strength. Air entrainment can make the mix stickier, so it may not flow as smoothly as concrete that has not been air entrained. This stickiness also may make troweling and finishing a little more difficult.

Magnesium or aluminum tools should be used to bullfloat and finish air-entrained concrete mixtures, because wooden floats tend to tear the surface of this type of mixture.

Recommended Air Content for Concrete Subject to Severe Exposure Conditions, by Aggregate Size*

Maximum-size coarse aggregate	Air content, percent by volume of concrete**
1 ½", 2, or 2 ½"	5 ± 1
¾" or 1"	6 ± 1
⅜" or ½"	7 ½ ± 1

*See text for formula variations and precautions.
**Air content in mortar alone should be about 9 percent.

Air Entrainment Variations and Precautions

- Add 2 percent to the recommended air-content values shown in the preceding table when structural-lightweight-aggregate concrete is being used.
- Because some entrained air tends to be lost in hot weather, the percentage of air should be increased slightly when placing concrete in hot weather.
- Certain water reducers, retarders, and high-range water reducers may affect air entrainment. Builders should consult their suppliers if they plan to use more than one admixture in the concrete batch.
- In concrete with a very low slump, the proper amount of entrained air is hard to attain without using a water-reducing admixture.
- Soupy or watery mixtures tend to lose entrained air rapidly when vibrated.

Concrete that is not air entrained may be used for basement slabs or wherever concrete is not exposed to freezing and thawing or deicers. However, since air-entrained concrete helps control bleeding and segregation and is easy to work, it may be preferred. If concrete without air entrainment is used, wood bullfloats and hand floats are recommended.

Air Entrainment

Air entrainment bubbles are microscopic spheres—only .0004 to .0400 inches in diameter. Air bubbles that you can see with your eyes are not air entrainment bubbles, but irregular-shaped bubbles of entrapped air. These bubbles are too large to help the durability of the concrete; in fact, they can weaken the concrete by forming channels that allow water to pass through the concrete as it cures. This condition is eliminated in concrete by proper consolidation through vibration of the fresh concrete mix.

Accelerators

Accelerators speed up the setting time and strength development of concrete and can be useful in cold weather. They can be combined with water-reducing admixtures. If used, proper attention must still be given to weather protection and curing because accelerators will not protect fresh concrete from freezing or too rapid drying. Always check manufacturers' recommendations before using an accelerator.

Calcium Chloride

Calcium chloride is a widely used accelerator; however, it has certain disadvantages and can pose problems if used in reinforced concrete because of potential damage to the reinforcing steel. It is not an antifreeze, but it speeds up the setting time and makes freezing damage less likely, especially if the concrete is insulated. Precautions to take when using calcium chloride include—

- Add calcium chloride in liquid form as part of the mixing water. If calcium chloride is added in dry form, it may not dissolve completely and can cause popouts and dark spots in the concrete as well as affect the air-entraining admixture.
- Never add more than 2 percent of calcium chloride by weight of cement. A greater amount causes the concrete to stiffen rapidly and makes placing and finishing difficult. Too much calcium chloride may also cause flash set, increase shrinkage (a cause of cracking), corrode reinforcement, and weaken and discolor the concrete.
- Do not use calcium chloride when the concrete is to contain embedded aluminum such as conduit. Serious corrosion can result, especially if the aluminum is near or in contact with steel and the environment is humid.
- Do not use calcium chloride if galvanized steel is to be permanently in contact with the concrete.
- Do not use calcium chloride if the concrete is to be exposed to soil or water that contains sulfates or is subject to alkali-aggregate reaction.
- Do not use polyethylene film for curing a concrete that contains calcium chloride because the concrete may become discolored if the film is wrinkled. Polyethylene film can be used successfully if the curing water is flooded onto the slab surface before the film is placed on top of the water.
- Some aggregates may not be compatible with calcium chloride. The concrete supplier should be consulted when there is any question at all about an aggregate.

Because of the potential for problems with the use of calcium chloride, many builders prefer to use other methods to accelerate the set of concrete. Nonchloride admixture accelerators now are widely available. Your local ready mixed concrete company should be able to provide you with information and advice on other accelerators.

Accelerating Concrete without Admixtures

Since calcium chloride requires such care and may increase drying shrinkage, the following alternative methods for accelerating the concrete should be strongly considered.

Type III, High-Early-Strength, Portland Cement. Type III portland cement gains strength almost twice as fast as Type I, normal, cement on the first day the concrete mixture is placed. However, the strengths are more or less equal after about three months.

Lower Water-Cement Ratio. A lower water-cement ratio accelerates strength gain, but using too much cement is costly and may cause cracking in the concrete.

Curing at a Higher Temperature. Higher temperatures speed up considerably the rate of strength gain, but the concrete may not retain its strength as well over the long term.

Heated Mixing Water. Warming the mix by using heated water can accelerate strength gain.

Retarders

Retarders, chemical agents that slow the setting time of the concrete, are useful in hot weather when regular concrete may set so quickly that it cannot be finished properly. Retarders are also useful when difficult placements require more time. Retarders that are also water reducers slow down the set while improving the workability of the concrete. Not all water reducers retard the set. Precautions to take when using retarders include—

- Retarders may entrain some air into the concrete. Check on the extent and allow for it when figuring the amount of air-entraining admixture.
- Slower strength gain usually occurs during the first one to three days when retarders are used. For this reason, form removal may need to be delayed.
- Other effects of retarders may vary with the product, so manufacturers' recommendations must be checked before using.
- A mixture may be accidentally retarded by even a slight amount of sugar in it. Such contamination can result in strength reduction and even failure. Sugar is only one of many possible contaminants that can have a deleterious effect on concrete. Making it a habit to use clean equipment and carefully monitor the way materials are handled on the jobsite can save you time and trouble in the long run.

Water Reducers

Water reducers, sometimes called plasticizers, make the concrete more workable with less water. Thus, concrete strength is increased (because the water-cement ratio is lower) and labor costs are reduced since the concrete is more workable.

Sugar in the Mix

Sugar can get into the concrete-mix water in the most unpredictable ways. Mix water stored in an old 55-gallon barrel at one jobsite was contaminated by sugar because the barrel had previously contained a syrup material. In another case, mix water drawn from a river at a point downstream from a factory contained sugar from factory wastes dumped in the river. Contaminants such as sugar can cause serious problems, even in small amounts. On the other hand, on more than one occasion a bag of sugar kept in the cab of a ready mix truck has been added to the concrete to keep it from setting and ruining the mixer when the truck became stuck in a muddy jobsite. Of course, the addition of the sugar ruins the batch of concrete—but the mixer is salvaged.

Some water reducers may increase drying shrinkage, and this may increase cracking. Other water reducers, however, will reduce shrinkage cracks, according to the manufacturers' claims. The most important factor affecting shrinkage is total water content per unit volume of concrete; therefore, reducing the amount of water required for workability improves concrete in several ways.

Water reducers may also be accelerators or retarders. The type chosen should fit the circumstance. For example, an accelerating water reducer is suitable for cold weather; a retarding water reducer is suitable for hot weather.

Some water reducers also entrain air. Allow for the extent of this entrainment when specifying amounts of air-entraining agent.

Water reducers are worth investigating since they may provide added strength without increasing labor costs.

High-Range Water Reducers

An ordinary water reducer can reduce water 10 to 15 percent while slightly increasing slump. A high-range water reducer (or superplasticizer) can reduce water up to 30 percent and can increase slump dramatically—from an original 3 inches to 7 or 8 inches. Since concrete mixtures with high-range water reducers are much easier to place, labor costs can be reduced.

A reduced-water concrete mix is easier to work not only because of the increased slump (called "flowing" concrete), but also because of the consistency of the concrete. Cement masons liken it to a temporary lubrication of the mix. It can be chuted more easily at a lower angle and is almost self-leveling.

The slump increase lasts about 30 to 60 minutes, depending on the product, and then the mix returns to its original low-slump condition. However, if a conventional water-reducing retarder is combined with a smaller amount of a high-range water reducer, the high slump can last as long as 2 or 3 hours. Combinations of admixtures should be field-tested with actual materials to verify that interaction does not create a problem.

If slump is lost too soon, more water reducer or more water may be added to retain slump, as long as the original mix design water-cement ratio is not exceeded. If exceeded, the additional water weakens the concrete. The return of slump to normal after a short time is not the same as the concrete setting up. Whether or not the high-range water reducer affects the set depends on the type and amount used.

Segregation of the large aggregate is usually not a problem with water-reduced mixes as it often is with conventional high-slump mixes. Mixes containing high-range water reducers that have extremely high slumps, however, may have some segregation.

Costs for high-range water reducers range from about $2 to $6 (in 1993) per cubic yard of concrete. However, the net concreting cost may be less because the cement content of the mix may be reduced and labor costs for placing and finishing may be decreased. Costs may be reduced even more by using a conventional water-reducing retarder along with a reduced amount of high-range water reducer. These steps also take some time pressure off the finisher.

Guidelines for Using High-Range Water Reducers

- Organize well and run a tight schedule. Before the mix arrives, have forms inspected, a cleared access for ready mix trucks, and a crew waiting.
- Use a conventional water-reducing retarder and less high-range water reducer to allow more working time and reduce cost. Add the water-reducer retarder before adding the high-range water reducer.
- Delay the final finishing until the concrete is firm enough to resist tearing. This delay may be longer when using a high-range water reducer than it would be for conventional concrete.
- If a ready mix truck is delayed and an 8-inch slump mix has to be placed against a reverted 3-inch slump mix, blend the old and new concrete together by vibration.

High-range water reducers often are used to improve placing, consolidation, and finishing of the concrete without adding water. Since special attention should be given to the type of cement, aggregate, and admixture and to the overall mix proportions, users should work closely with reputable suppliers, check manufacturers' recommendations, and seek advice and publications from professional organizations. Addresses of several helpful organizations are listed under "Sources of Additional Information" at the back of this book.

Finishing

Poor finishing can seriously impair the concrete surface. For good finishing, one of the most important skills is patience. The mason must not do anything to the concrete while bleed water is on the surface.

When concrete is worked while bleed water is present, the sand, cement, and water at the surface are mixed together into a thin surface paste with a high water-cement ratio and a high sand content at the concrete's most vulnerable point. Crazing, dusting, and scaling or spalling are almost certain to result.

Good finishing done at the right time gives the concrete a hard dense surface, making it more durable and impermeable. For more information on finishing concrete, see Chapter 4.

Uniformity

All areas of the concrete should be worked on uniformly and methodically. The following rules on uniformity should be strictly adhered to—

- Concrete should have about the same wetness all over when it is being cured, and the cure should be tapered off gradually, instead of shocking the concrete by sudden drying.
- The subgrade should be uniformly damp when the concrete is placed.
- If the concrete is heated during curing, the heat should be uniformly distributed and tapered off gradually (maximum reduction of 50 degrees F over 24 hours) at the end of curing.
- The mix should be uniform from batch to batch and should not be worked so forcefully that the aggregate becomes unevenly distributed.

Concreting in Extreme Weather

Fresh concrete loses moisture much faster in hot, dry, or windy weather. Drying too fast can cause the concrete to be less strong and less durable. Cold weather is also a dangerous time for freshly placed concrete; for if the concrete freezes before it sets, it will be less strong, durable, and watertight. Good curing can help reclaim some but not all of the desirable properties. A hard rain can ruin the surface of freshly placed concrete. Therefore, it is vitally important to make good preparations for extremes in weather.

Guidelines for preparing the subgrade in extremely hot and cold weather are included in Chapter 2. For a more detailed discussion of placing and finishing concrete in hot, dry, or windy weather and in cold weather, see Chapter 4. Measures that apply to curing concrete in extreme weather are discussed in Chapter 5.

Controlling Cracking

Concrete expands and contracts slightly with changes in its moisture content and temperature. Since it is low in tensile strength, it is apt to crack if it is restrained from moving as its volume changes. Concrete expands and contracts from temperature and shrinkage as much as $\frac{1}{16}$ inch in 10 feet. The problem is not to eliminate cracking, but rather to—

- Minimize cracking by using a minimum amount of water; using good curing techniques; preparing a good subgrade or structural support; using isolation joints to separate differing concrete elements, if appropriate; and avoiding abrupt changes in temperature and moisture, particularly during curing.
- Control cracking by proper jointing or use of reinforcement or both.

Detailed information on jointing, curing, and reinforcement to control cracking is presented in Chapter 5.

Poor or Nonexistent Jointing Causes Cracking

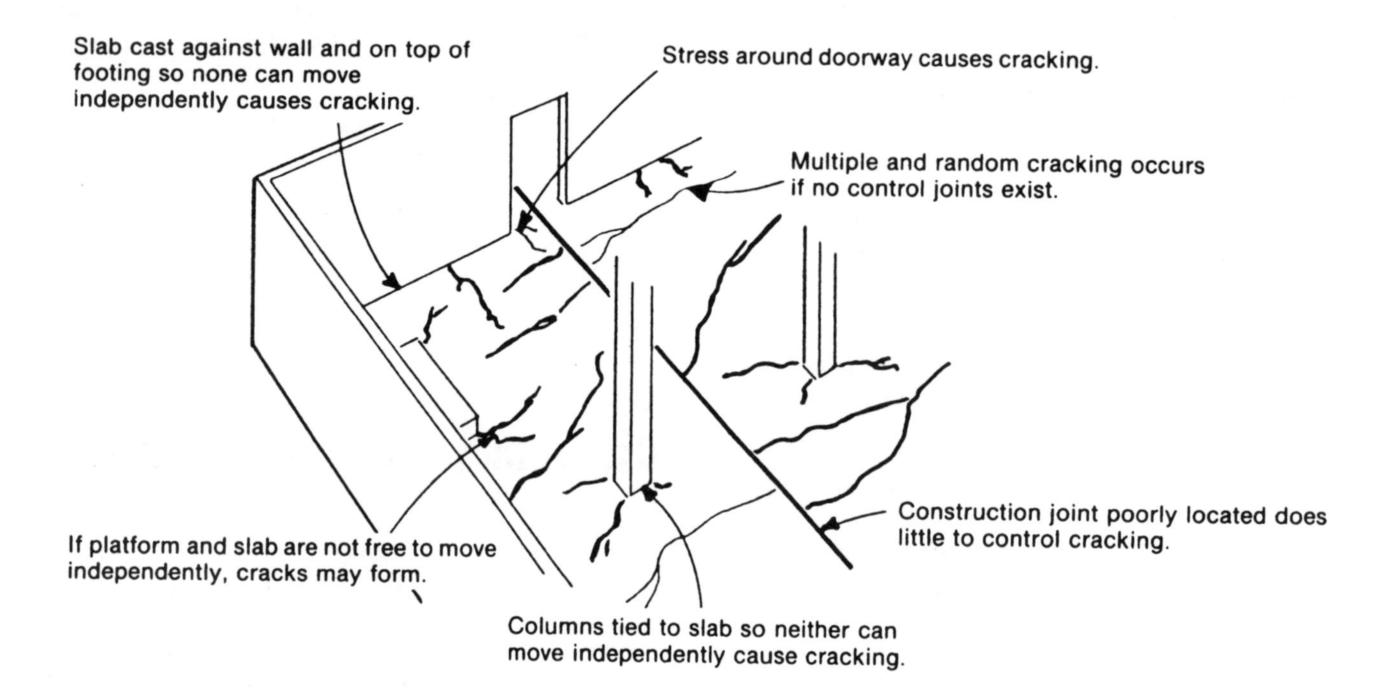

Controlling Cracking with Good Jointing

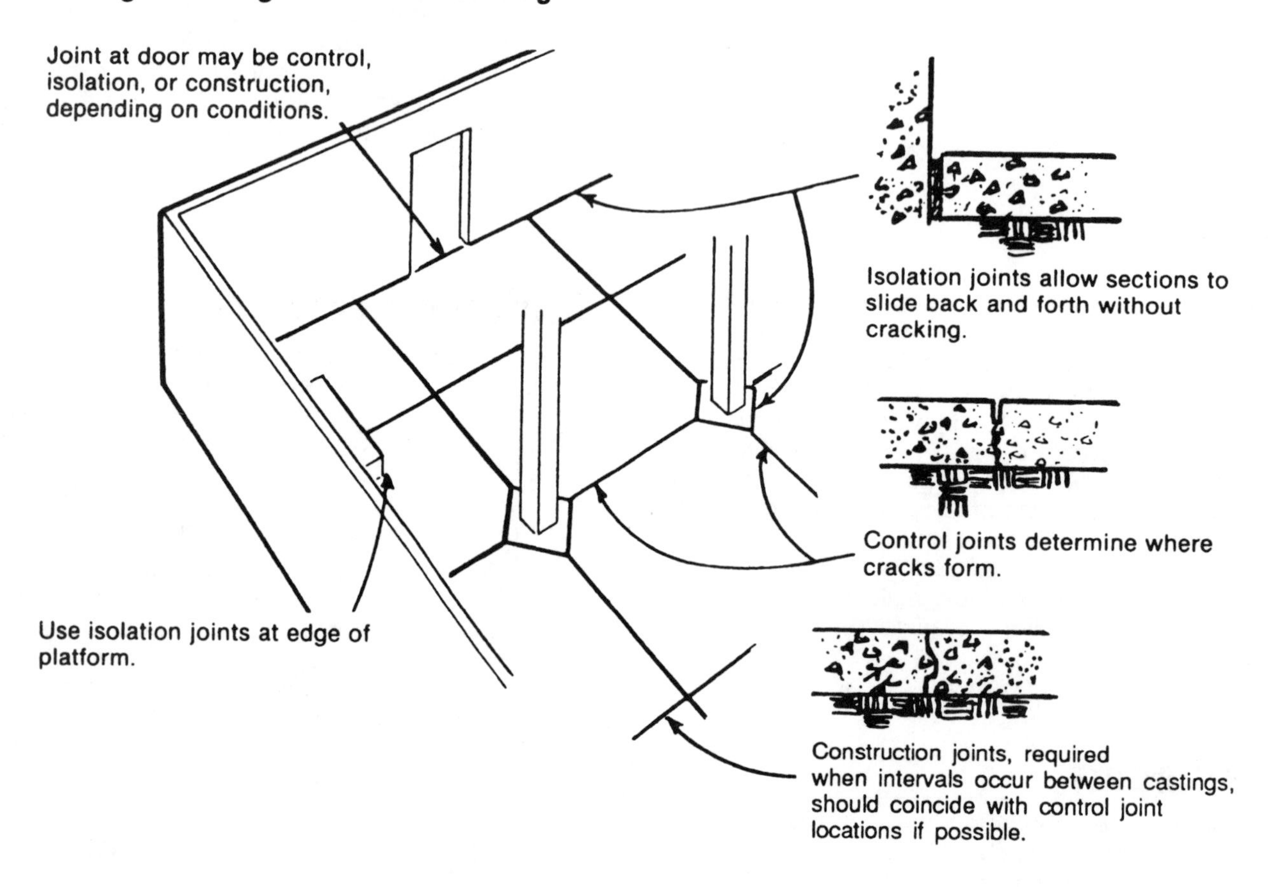

Wire Mesh

Mesh is often placed in slabs to control cracking. It distributes stresses so that many fine cracks develop rather than a few wide ones, and it helps hold the cracks tightly closed. The mesh keeps the aggregate interlocked and thus helps prevent differential settlement.

Differential settlement occurs when the concrete on one side of a crack drops lower than the other side. In exterior slabs such as sidewalks, patios, and driveways, preventing differential settlement can be particularly important. Insured warranty programs such as the NAHB Home Owner's Warranty program (HOW) may outline standards for maximum vertical displacement for basement and garage slabs. Cracks in slabs that have a finish floor covering should be repaired if the cracks will rupture the finish material or are readily apparent.

Debate continues about the value of mesh in slabs on grade. However, since codes may require it, the question is often how to best use it rather than whether to use it.

Slab control joints can be spaced farther apart when mesh is used. The lap of the mesh should not coincide with a control joint.

Wire mesh should be located 2 inches down from the top of the slab to control surface shrinkage cracks.

Fiber Reinforcement

Fibers made from synthetic materials or metal also can be added to the concrete to help reduce shrinkage cracking. However, fibers are not substitutes for primary structural reinforcement. Because the fibers are not oriented specifically within the concrete, they are inefficient in providing added strength; furthermore, they are not used in sufficient volume to replace steel rebars. Unlike structural reinforcements, fibers primarily provide benefits when the concrete is still plastic.

Fibers are most commonly added to concrete for slab-on-grade construction to reduce early plastic shrinkage in fresh concrete and to increase impact- and abrasion-resistance and toughness.

Reinforcing Steel

Reinforcing steel (sometimes called rebar) typically is not used in footings or slabs on ground unless it is called for in the plans, specifications, or building code.

Two #4 (½ inch) rebars often are used in the footing to prevent the footing from settling at one or more points.

Special Cement

Shrinkage-compensating concretes made with expansive cement have been used to reduce cracking, but to date these have been used mostly in commercial construction. The concrete actually expands as it cures, putting the rebars under tension. Slabs can be as large as 100 feet between crack-control joints with this special cement. Carefully placed reinforcements in the slab contain the expansion and put the concrete in slight compression. This compression helps prevent cracking and curling of the slab. For more detailed information on this subject check with suppliers or industry experts.

Earthquake Considerations

Areas of the United States that require consideration for major earthquakes include California, Washington State, South Carolina, and the Mississippi Valley. However, virtually all areas of the country experience slight to moderate earthquakes from time to time. Earthquake resistance is highly dependent on the foundation of a house; therefore, concrete foundations must be properly constructed and reinforced to provide appropriate resistance. Especially important are the following points—

- Foundation walls and spread footings should be reinforced for earthquake resistance.
- Footings should have uniform soil support.
- Reinforcing bars should have adequate lap splicing as required by the building code.
- Sill plates should have anchor bolts spaced at no more than 6-foot intervals.
- Concrete slab-on-grade foundations should be reinforced to act as a horizontal floor diaphragms.

Summary of Basic Requirements

To make concrete strong and hard, crack and spall resistant, weather durable, and waterproof—

- Start with a firm, uniform subgrade.
- Use a minimum amount of water in the mix. The slump should be low. If necessary, use water reducers or high-range water reducers to make low-slump concrete more workable.
- Use air entrainment for exterior concrete or any other concrete exposed to freeze-thaw or deicers.
- Work with your supplier to be sure the mix design will meet all of your job requirements.
- Place and work the concrete in a way that avoids segregation of the mix.
- Finish the concrete correctly: do not do anything when bleed water is present (to avoid dusting, scaling, and crazing); make control joints one-quarter of the slab depth; make isolation joints where appropriate; float and trowel for a hard, dense surface.
- Become knowledgeable about how accelerators, retarders, sealers, and other admixtures may interact to affect the quality of the concrete.
- Adjust your methods appropriately to compensate for extremely hot or cold weather.
- Cure the concrete properly.

Chapter 2

Building Siting and Layout

Siting of the house on the building plot is an important step. Building and zoning regulations for the city or county must be met. Consideration should also be given to location of underground utilities, topography, rainwater runoff, elevation relative to the street, and groundwater and soil conditions. To start construction of the concrete footings, foundations, and slabs without adequate concern for the proper siting would be unwise.

The first step is to contact the local building authority for the building site. In order to obtain a building permit, house plans including a plot plan will be required. If the house is a custom design, the architect will provide the plot plan. If the house plan is a conventional design without specific siting details, then a plot plan drawn to scale will be needed.

Plot Plans

A plot plan shows essential information to locate the house on the lot, to protect underground utilities, and to set foundation elevations. Information that should be provided includes the following—

- Legal description locating the lot.
- Property lines and dimensions.
- Location and dimensions of the proposed building.
- Location of water, sewer, electric, and gas lines.
- Location and dimensions of walks, driveway, steps, and patio.
- Location and dimensions of easements.
- Elevations of first floor, garage floor, driveway, street surface, and finished grade at each corner of the building.
- Location and size of existing trees and other features of the lot.
- North compass direction and, if possible, a permanent benchmark that is usually marked with elevation of 100 feet.

Additional information useful in layout of the building may be existing and proposed soil contours, temporary electric and water connections, and soil survey data.

Location of Utilities

The locations of all buried utility lines should be marked on the site. Digging around buried utility lines can result in damage to the lines and also can be dangerous—particularly in the case of gas lines. In some localities the law requires a call to the utility company before any digging can begin. Utility companies usually are happy to locate and mark their installations in order to prevent possible damage.

Doublecheck Utility Line Locations
Although an existing plot plan may show buried utilities, changes may have been made in the location since the last drawing was made. For example, one builder trusted the location of a sewer line shown on the drawing. When he drilled deep piers to support the house foundation, he hit something hard but did not realize that it was the sewer. Later when the neighbors complained to city building officials that their sewers were backing up, it was discovered that the piers had damaged the sewer. Expensive reconstruction was required to correct the problem.

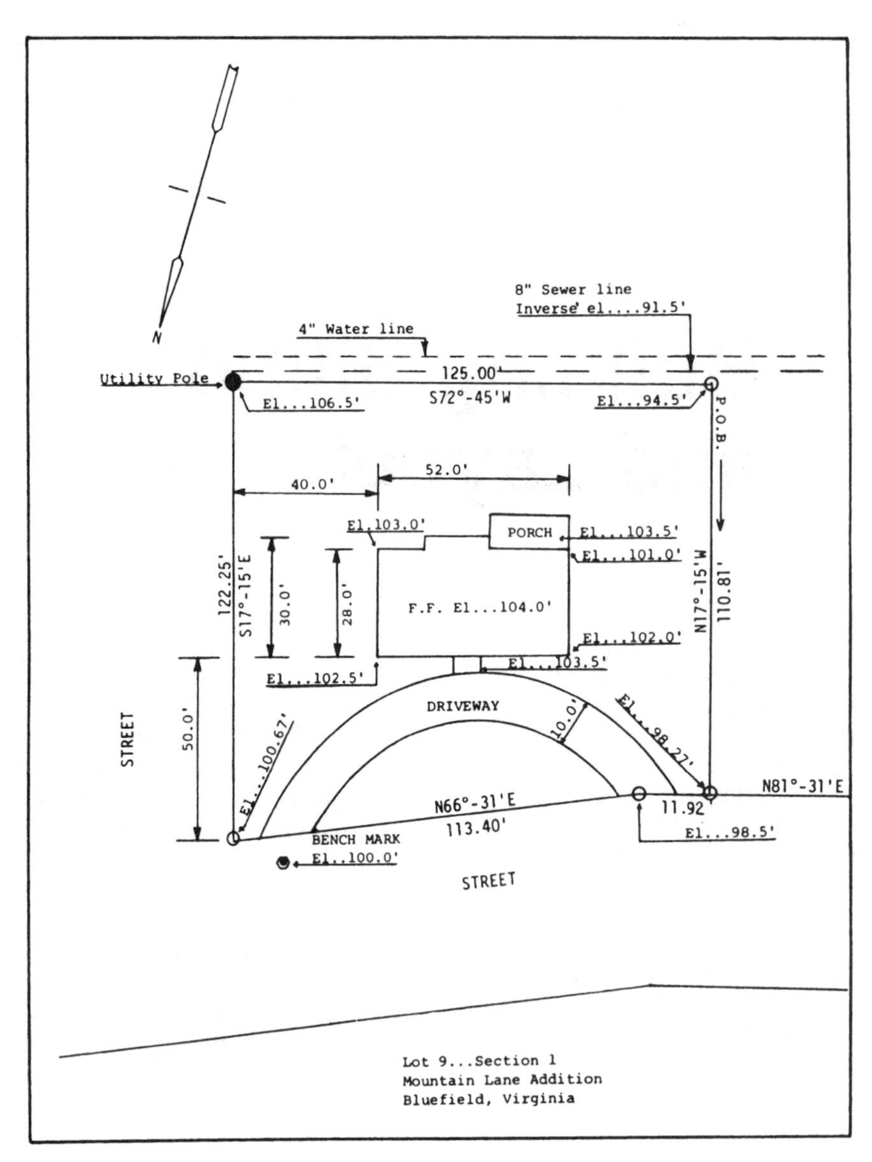

Plot plan

Making a Plot Plan

If a plot plan is not provided with the house drawings, one must be developed with the needed information and a drawing made so that subcontractors, inspectors, and workers can properly locate and start the foundation. The up-front time and effort spent on planning can be valuable in preventing errors, scheduling the work, and keeping the project within budget.

A survey must be made to locate property lines. A legal description of the lot and the location of lot boundaries may be obtained from the city or county property recording office. Depending on local building and zoning requirements for setbacks of the house from lot boundaries, draw the house and appurtenances to scale on the plot plan, as in the example. Determine the excavation needed to start the foundation.

Building Permits

Any major construction—including building or remodeling a house, garage, driveway, sidewalk, patio, or other structure—requires a building permit. Permits are required to ensure that work is done in accordance with state and local building codes. Depending on the type of construction, inspections of the work at critical steps will be required. For example, after the footing or foundation is placed, the inspector will check for proper elevations and setbacks from property lines. The inspector also can give advice about the building layout.

Layout and Site Preparation

Based on the plot plan, stake out the house foundation on the site and prepare to dig the footings. Some builders rely on the owner or the survey engineer to provide the proper structure location in relation to setbacks or property boundaries. Clarifying who is responsible for the precise location of the building on the site *before* excavation begins can help prevent potentially costly problems later on. One method is to set batter boards that indicate the house location and provide a reference for elevations.

Setting Batter Boards

Batter boards are temporary wooden stakes and horizontal boards erected 3 or 4 feet outside each corner of the house. These boards are set at a level reference elevation. The boards are marked to locate the foundation line. Nylon strings are then stretched taut between the batter boards to indicate the location of the foundation.

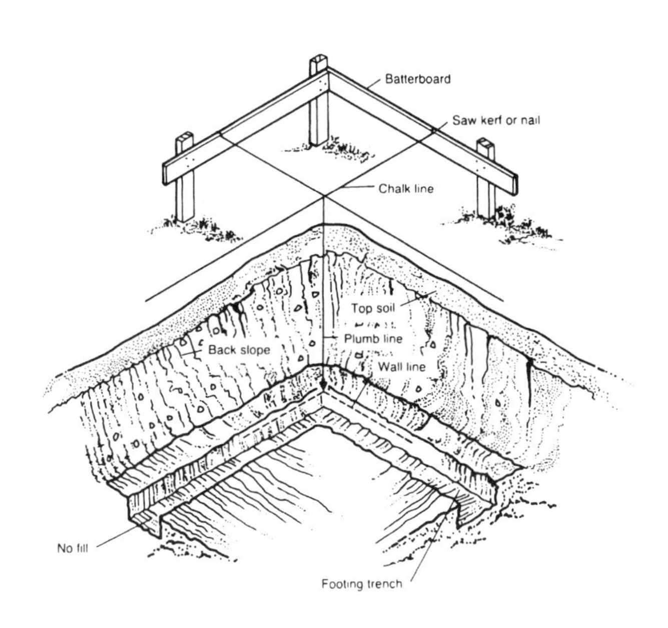

Setting Elevations

Because the batter boards are all on the same level, measuring down from the strings shows how much excavation is necessary for the footings. The strings may be removed temporarily to do the digging. After each major activity such as excavation, placing footings, and setting foundation forms, the elevations should be checked with the batter boards and string. Alignment of the batter boards and strings also may need to be verified after each activity because they may be disturbed by the work.

Builder's level

Using a Builder's Level

A builder's level may be used to set elevations without batter boards. The builder's level is a simple instrument to set elevations, establish lines, and turn angles. It cannot measure vertical angles as a surveyor's or engineer's transit can, but it can do most jobs needed for house construction. Elevations are usually given in hundredths of a foot, rather than fractions of inches (as on a carpenter's rule); therefore, you may have to calculate mathematical conversions to locate various levels. Using a combination transit with optical plummet and laser level can offer an advantage over batter boards because the laser is unaffected by the excavation or construction work.

Excavations

Obviously, the foundation of a house must rest on sound soil. Local building codes usually specify the minimum bearing capacity and acceptable condition of the foundation soil. Footings should be deep enough to reach soil that does not contain organic silt, roots, soft clay, or other unsatisfactory materials. If poor soil support is expected, drilled piers or deep foundations may be needed. If this is the case, consult with an experienced engineer familiar with these specialized techniques.

Footings may be either excavated into firm soil or they may require forming if the soil cannot support itself. If excavations are not precise enough (with about a 2-inch tolerance), side forms may be required. Minimum width and depth of the footings is usually set by the local code.

Sometimes on a sloping site, the footing bottom may need to be placed on fill dirt, or the footing may be stepped. Any fill dirt or disturbed soil must be carefully compacted. Low-strength concrete fill (flowable fill) may be used to fill areas under the footings. In some cases, piers must be drilled to sound soil below the fill material. Footings resting on concrete piers will need reinforcing bars to span between the pier supports. Consultation with a professional engineer may be necessary.

Footings must rest on soil that is below the frost line for the building site. This depth ranges from 12 inches in the South-Central states to 4 feet in the Great Lakes states. Furthermore, concrete should never be cast on frozen soil.

Excavations for walks, stairs, patio slabs, and driveways are similar to those for footings, but may involve special considerations. With topsoil removed from the area to be concreted, prepare the subgrade. This is a most essential step for any concrete slab on ground. Serious cracks, slab settlement, and structural failure can very often be traced to a poor subgrade. The subgrade should be uniform, reasonably hard, free from foreign matter, and well drained. Undisturbed soil supports a concrete slab better than soil that has been dug out and replaced with poorly compacted fill. For a small job, hand tampers may be used; however, for large-volume work, mechanical rollers or vibratory compactors are recommended.

Setting footing and foundation forms is the next step in the house construction.

Layout Guidelines for Concrete Slabs

Recommended Slopes

Driveways, Patios, and Sidewalks. Always slope a driveway so that water runs away from the house. A slope may need to be modified to be sure it is compatible with the earth slope. The minimum slope recommended on a concrete surface is ⅛-inch drop per foot. Grass slopes adjoining concrete driveways should have a minimum slope of ½-inch drop per foot to ensure proper drainage. Some special concrete surfaces may require greater slopes to ensure proper water runoff after rains.

Basement Slabs. Basement slabs should slope at least ⅛ inch per foot toward basement doors, sumps, or drains.

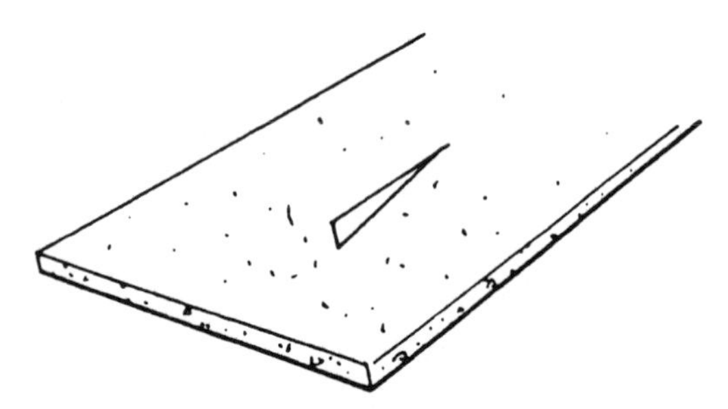

The slope may be toward the street

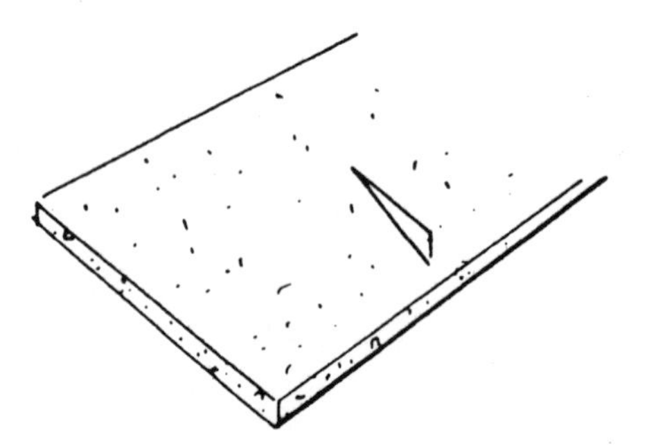

from one side to the other

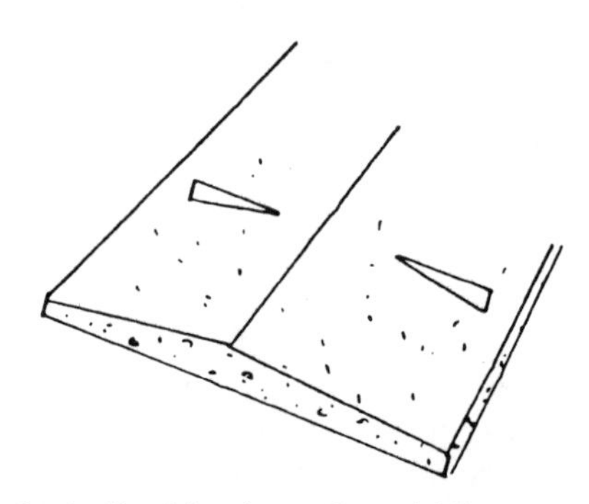

to both sides from the middle

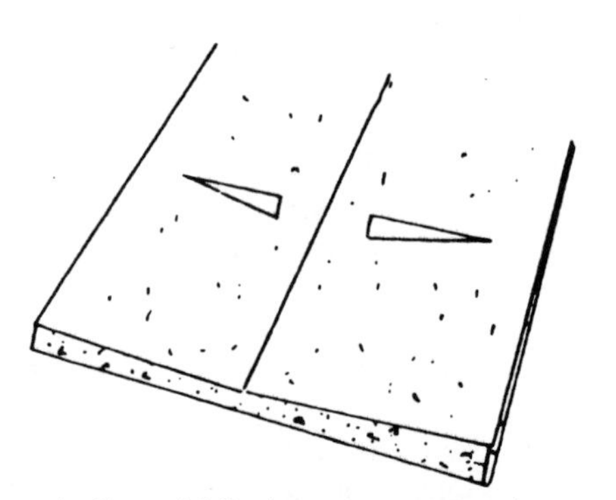

to the middle from both sides

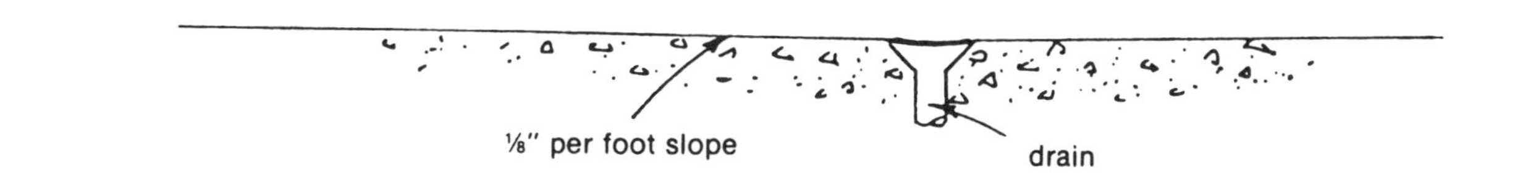

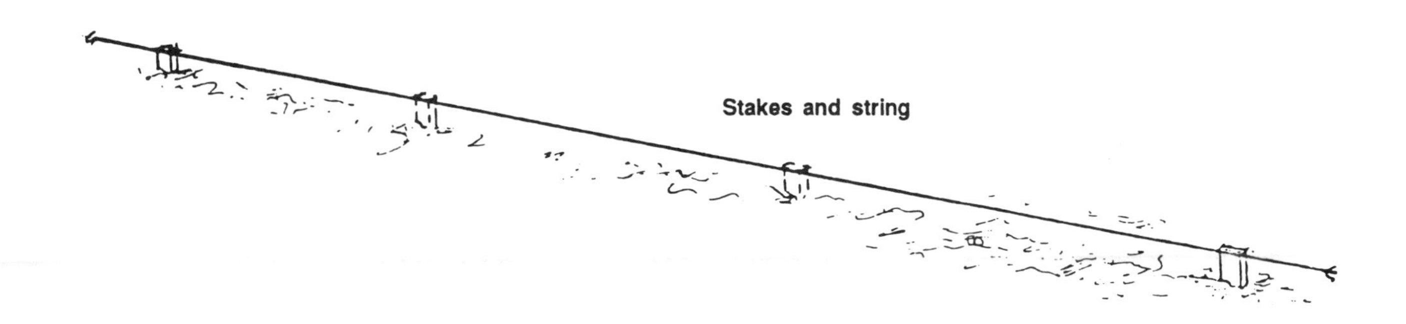

Stakes and string

Recommended Thicknesses

Driveways. If a driveway will be used by cars only, the concrete slab should be 3 ½ to 4 inches thick. If it will be used by both cars and trucks, the slab should be 4 ½ to 5 ½ inches thick.

Patios, Floor Slabs, and Sidewalks. Slabs for patios, sidewalks, and floors (including basement and garage floor slabs) should be 3 ½ to 4 ½ inches thick. Wire mesh reinforcement may be advisable to minimize potential cracks.

A lesser thickness of about 3 inches may be satisfactory if (1) the subgrade is firm and uniform, (2) the concrete has a low water-cement ratio and is well-cured, and (3) it is not to receive heavy loads. Always check code requirements for minimum thicknesses.

Setting the Grade Using a Builder's Level

Set the end stakes with a builder's level, allowing for slope; then stretch a string for intermediate stakes.

Setting the Grade Using a Story Pole

Using a story pole (a stick marked or cut to the right length) is a good technique for basement slabs if the overhead joists are in place. Measure down from these joists with the pole and set stakes around the entire slab area. Be sure to allow slope for drainage.

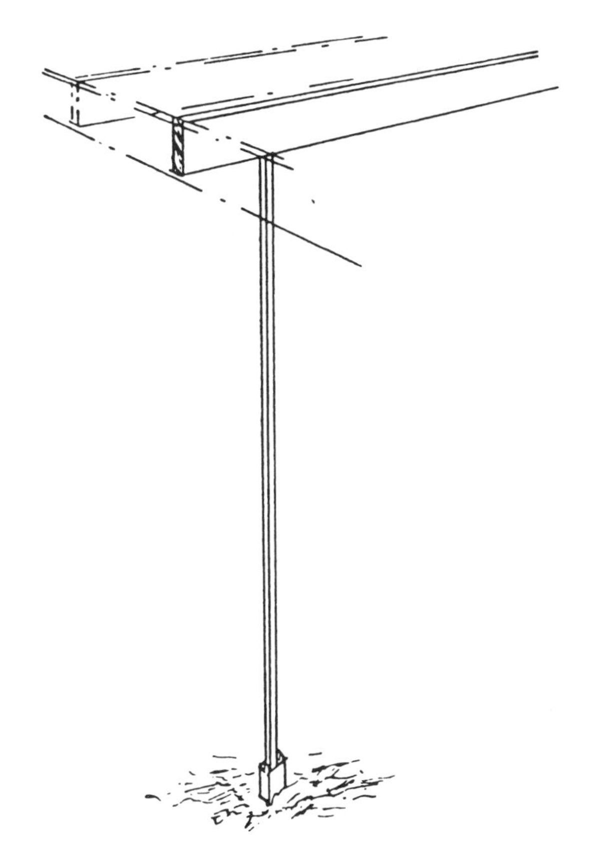

Story pole

Preparing the Subgrade

The basic rule for preparing a subgrade is to keep it uniform in—

- firmness,
- grade, and
- dampness.

If the subgrade is not uniform, the concrete will be under more stress and may develop cracks.

Fill

Sometimes granular fill is used as a subbase or to fill low spots. Use sand, crushed stone, gravel, or slag. Do not use cinders, clay, or vegetation such as leaves, because they will not give firm and uniform support. The fill should have the same slope as the finished concrete. Earth fill that is used for low spots should match the type and moisture content of the surrounding soil. All fills should be granular and well tamped. Hand tampers are inexpensive and are available from tool companies. Vibratory compactors are sometimes helpful for larger jobs. Low-strength concrete fill (flowable fill), which is self-compacting and self-leveling, can also be used.

Dampening

Dampen the subgrade before placing concrete. Dampening keeps moisture from being drawn too rapidly from the concrete and helps avoid cracks and discoloration.

If possible, dampen the subgrade the night before or at least long enough before placing the concrete so that the subgrade is uniformly damp with no standing water when the concrete is placed.

The Subgrade in Hot Weather

In hot weather, dampen the subgrade well in advance of concrete placement. In some hot and arid parts of the country, workers dampen the subgrade for 24 hours before placing the concrete. More typically, the dampening is done the night before. The water should be well absorbed into the earth, with no standing water remaining at the time of concrete placement. The forms may need wetting too, especially if they are not treated with a release agent.

The Subgrade in Cold Weather

Do not place concrete on a frozen subgrade. A frozen subgrade is unstable and may either heave or settle unevenly, and the concrete will crack. Some builders avoid frozen subgrades by excavating the last foot or so the night before placing. Others may place straw or insulating blankets over the subgrade to prevent freezing. Still others may close in a basement, possibly even heating it, before placing concrete.

Preventing Earth Collapse

Sandy soil. Trenches or basements dug in sandy earth are potentially dangerous. Even if the soil stands for some time after a vertical cut is made, sandy soil may fall at any time without warning. Sandy soil is in greater danger of collapse from a moisture change such as rain or a change in soil moisture, including shifts in the water table (the level of free water in the ground). Sandy soil also may collapse as a result of vibrations from nearby blasting, traffic, or equipment or from overloads of materials, equipment, or workers standing near the cut.

Clay soil. Clay soil usually can stand a small vertical cut. However, clay can be dangerous if the cut is over 4 feet deep or if it is soft, regardless of the amount of moisture present. Be aware of OSHA regulations about where persons are allowed to stand while working an excavation. If a 2x4 can be rammed into the clay, beware of a collapse. Added moisture will increase the danger.

Silty soil. Silt has characteristics of clay soil, but usually is found mixed with sands; its resistance to collapse is uncertain, so it is safest to treat it like sandy earth.

Loose fill of any type of soil poses a danger of collapse if it is sloped more than 45 degrees.

Unstable Soils

Soils in some locations, especially in Texas, Oklahoma, Colorado, and New Mexico, may be so unstable that they require special foundation construction. Unstable soils often shrink or swell under the corners of a house, and the stress applied to the foundation by these soil changes can cause cracks to form in the masonry walls. Cracks caused by shrinkage usually are very small at the foundation, but widen near the house eaves. Normal use of rebars in the footing or grade beam usually will not prevent these cracks. Special techniques, such as post-tensioned slabs and grade beam-piers, have been developed for building on such soils. It may be necessary to consult a professional engineer if you are dealing with these types of conditions.

Pre-Treatment for Termites

Termites are a moderate to heavy hazard over most areas of the United States. Protection for wood structures can be provided by chemical barriers and by physical barriers such as metal shields. For concrete slab-on-grade construction, chemical treatment of the soil and use of pressure-treated bottom plates may be required to effectively control termites. Clearance of all wood above soil should be at least 6 inches. When an under-floor crawl space is built, floor joists should be 18 inches above the soil.

Radon Considerations

Soil gases such as radon are a problem in many areas of the United States. Often the presence of radon gas may not be identified until after a new home has been built and has stood on the site for a while. Where precautions to prevent entry of radon gas are desired or advisable, preventive design and construction measures can be incorporated into the home's plan. The Environmental Protection Agency recommends that protective measures be taken if the average annual level of radon is higher than 4 picoCuries of gas per liter of air.

Protective measures generally involve one or both of the following: efforts to prevent entry of gases into the home, and provisions for venting gases that do penetrate in order to prevent a buildup to high concentrations. A highly effective protective measure, subslab depressurization systems use passively vented methods (or, if needed, a fan-driven vent system) to remove radon concentrations from beneath the slab. In order for the gases to reach the piping system, they must flow through a gas-permeable layer such as a granular subbase (which also provides a good transfer of slab loads to the subgrade). In this system, the gas-permeable layer should be topped by a soil-gas retarder (an impermeable barrier such as 6-mil reinforced plastic) that prevents infiltration of the gas-permeable layer by concrete when the slab is placed. The soil-gas retarder should be continuous under the slab—it will work effectively only if *no* tears or perforations occur during installation—and overlapped a minimum of 12 inches at the joints.

Approximately 3 inches of a damp sand or 2 inches of gravel should be placed on top of the soil-gas retarder to help prevent plastic-shrinkage cracks from developing in the basement slab. A vent stack is inserted into the gas-permeable layer, using a T-fitting to prevent closure of the stack if movement of the stack occurs. As with all joints in the basement area, the joints between the bent stack and the slab should be closed with a durable, compatible sealant material.

If possible, avoid placing forced-air heating or cooling vents within the gas-permeable layer. If these lines must be placed in the gas-permeable layer, be sure they are constructed with air-tight joints and protected against corrosion so that radon will not penetrate the lines and thereby gain access to the residence.

Cost Estimating for Foundations and Slabs

Cost estimating for foundations and slabs includes material and labor costs for excavation or fill, soil compaction, footings, foundation walls, piers and pedestals, vapor barrier (if used), gravel or sand base, slabs, equipment, rebars, mesh, anchor bolts, termite protection, and backfill. Some builders list costs for batter boards under the building layout rather than under foundations and slabs.

Detailed quantities of each item are calculated and entered into a foundation take-off sheet. Labor costs are entered, based on historical company data or on standards such as Dodge's or Means estimating tables. Allowance is made for building location, anticipated weather conditions, site conditions, and other factors. Of course, labor costs are the most variable of all costs in the estimate. Careful and continuous project management is the best way to keep costs under control.

Summary of Layout and Subgrade Preparation

Proper preparation prior to form erection can be verified by checking the following list—

- Check that the building is located on the lot as shown on the plot plan and that it conforms to local codes.
- Check to be sure all required permits have been obtained.
- Check that the foundation and slabs have the correct elevation and slope for adequate surface water runoff.
- Ascertain the location of all underground utility lines.
- Check that the subgrade is well compacted and at the proper elevation.
- If specified, the sand or gravel base for slabs should be ready.

Chapter 3

Forming

Formwork encompasses the forms, supporting members, hardware, and all necessary bracing to contain the wet concrete mix. Forms serve to contain the mix in a specific shape until the internal strength of the concrete will hold that shape. This chapter includes suggestions and instructions for forming footings, curbs, gutters, slabs, straight walls, and stairs.

Saving Dollars on Forms
Economy is achieved in forms and forming by—

- Using form lumber several times.
- Using a fast and easy method of building forms.
- Using a fast and easy method of form removal.
- Cleaning and applying a release agent for reuse of forms.
- Using standardized modular forms when the forms may be reused many times.
- Using resin-overlay (high-density overlaid) plywood for reusable forms.

The basic rule for forming is to think ahead. Design the forms for the least cost overall by considering the reusability of materials and the labor both to build and disassemble them.

Footings

A footing must have firm and uniform support. Therefore, an excavation that has been dug too deep or too wide should be left that way and the extra cut filled with concrete. This may be more economical than adding and compacting fill.

Excavated footing

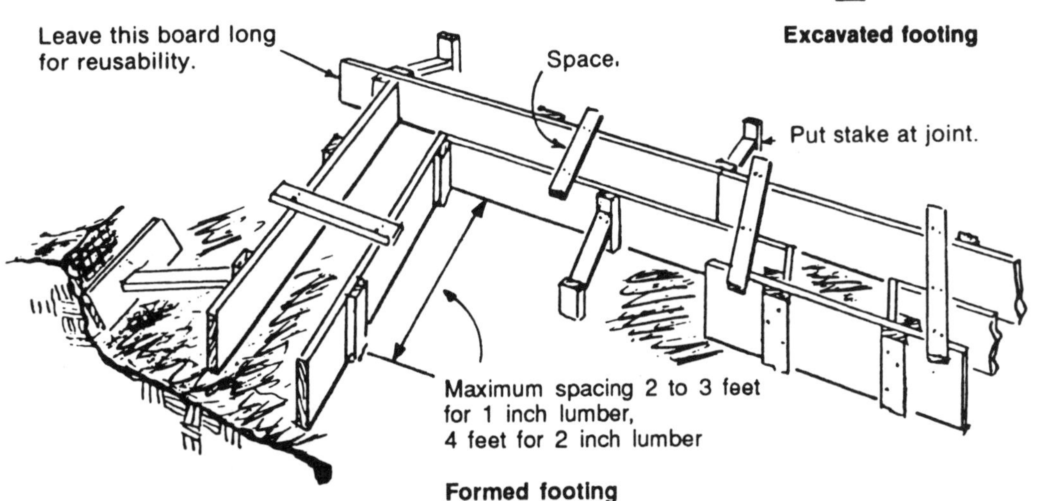

Formed footing

Metal Stakes and Spreaders

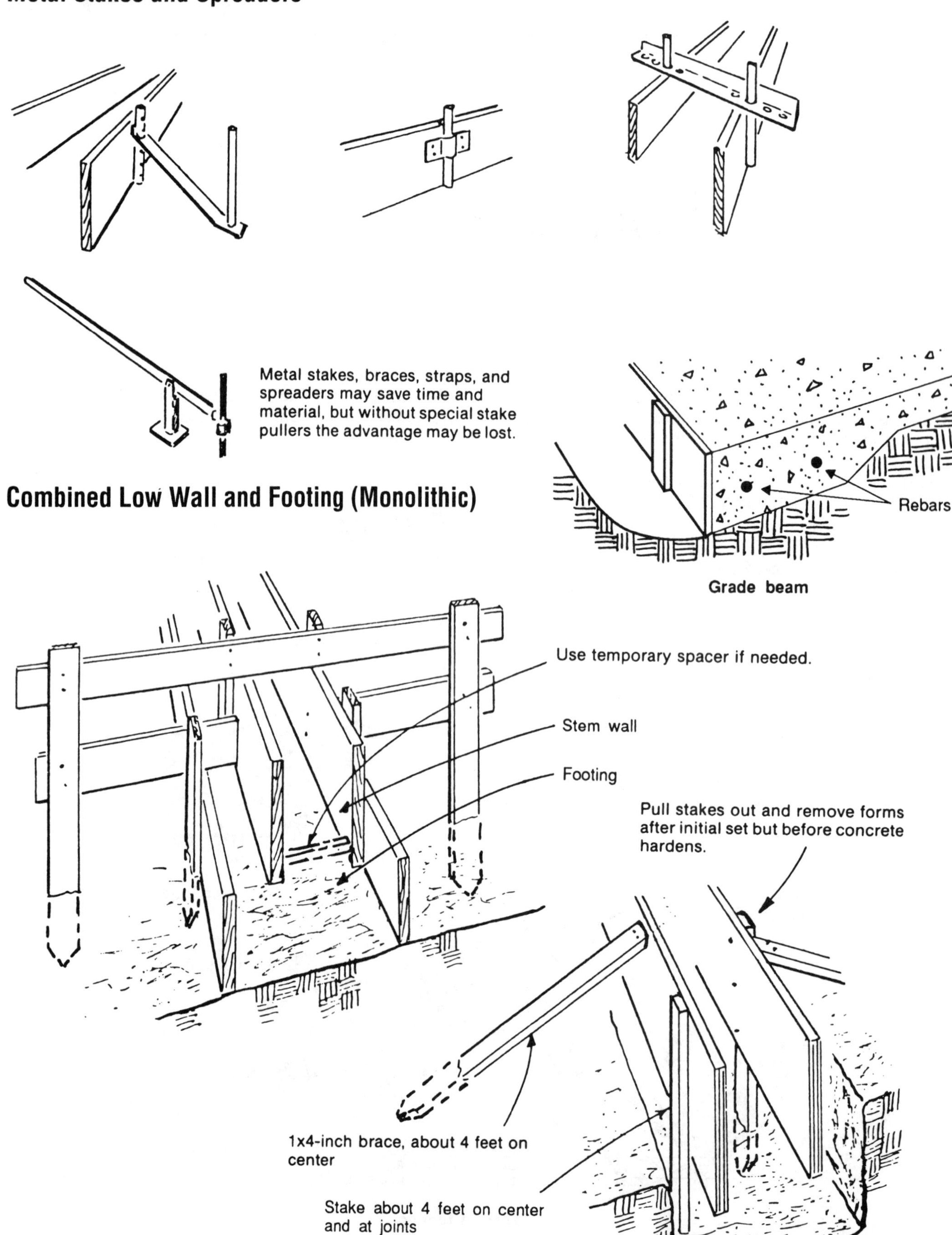

Stepped Footing

Stepped footings are sometimes used to follow the ground line on a sloped building site. Adequate strength at the step requires that the horizontal overlap between steps not be less than 2 feet. They should be placed monolithically or tied together with reinforcing bars.

If a masonry wall is to be placed on a stepped footing, the footing should fit with masonry courses.

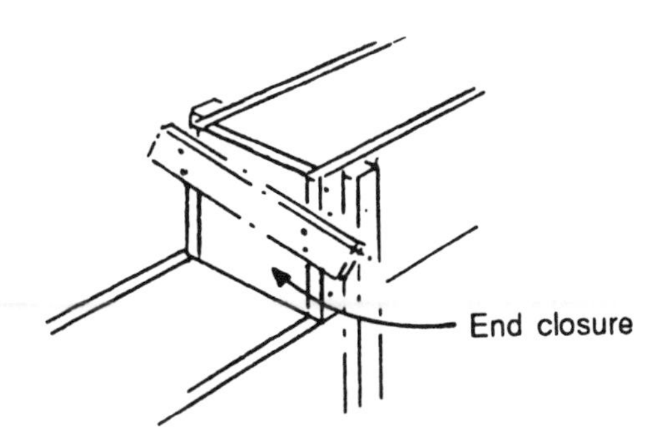

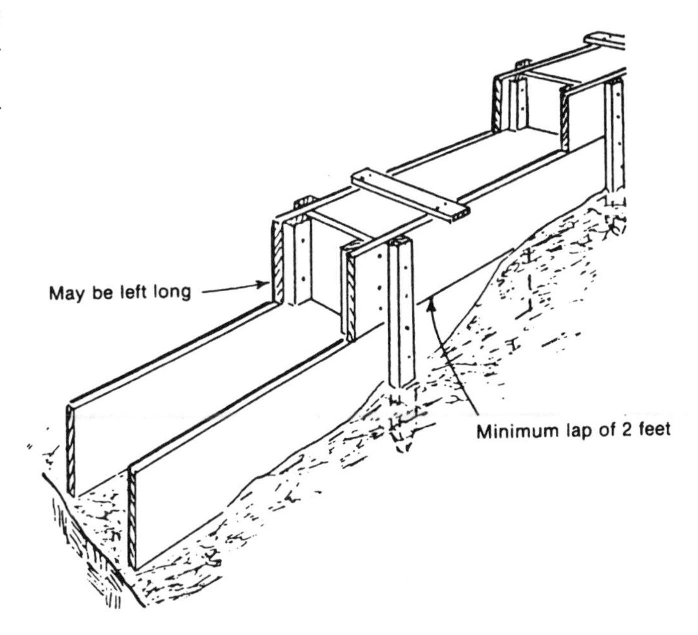

Corner Fastenings

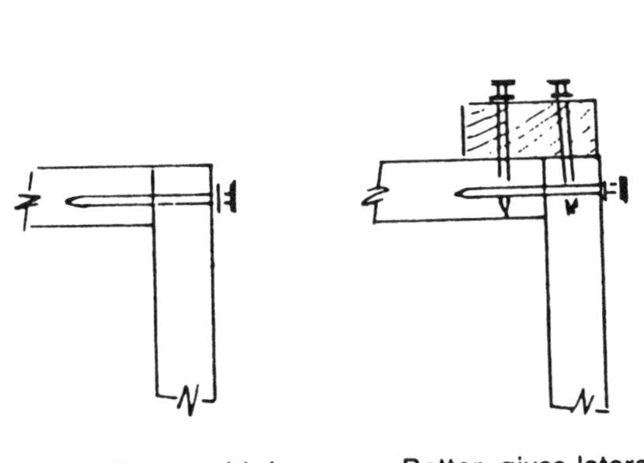

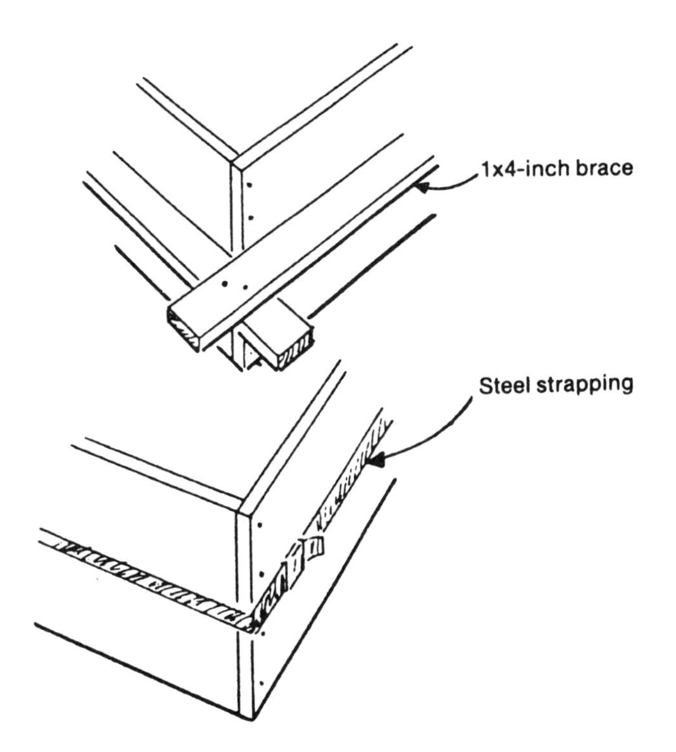

Forming Curbs and Gutters

Curbs can be finish-shaped with a darby or hand float. Driveway entrance curbs may not require a template. Sometimes depressions for entrance curbs can be shaped entirely with a cement finisher's trowel if the concrete has low slump.

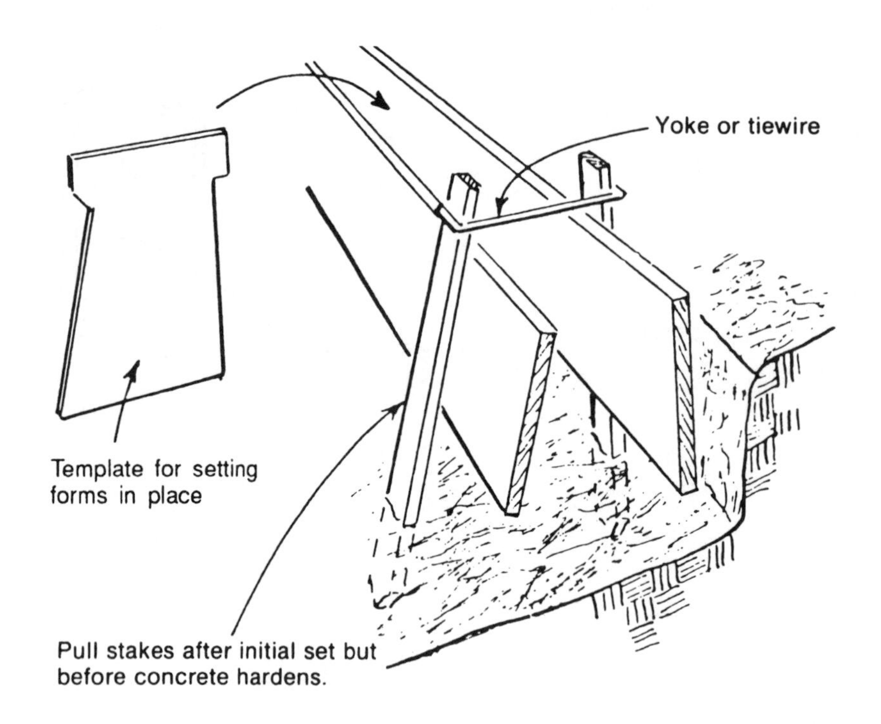

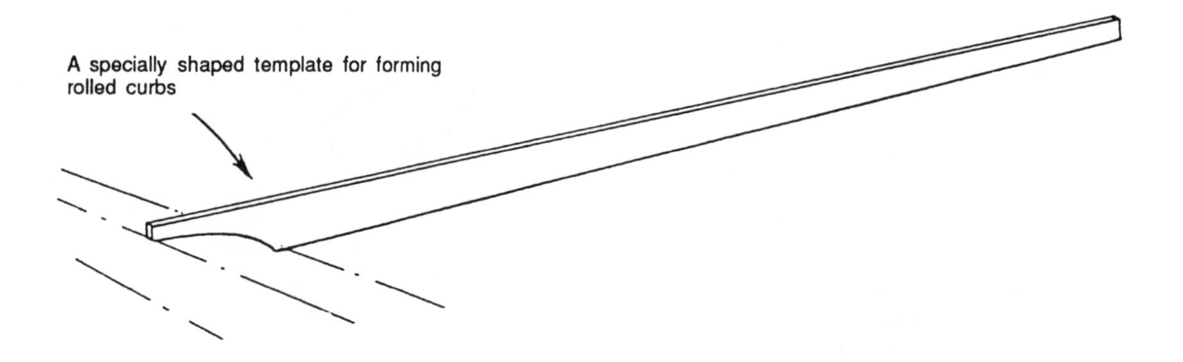

Forming a Slab on Grade

After the grade is marked on the end stakes, stretch a line for setting the form.

Put stake at all butt joints.

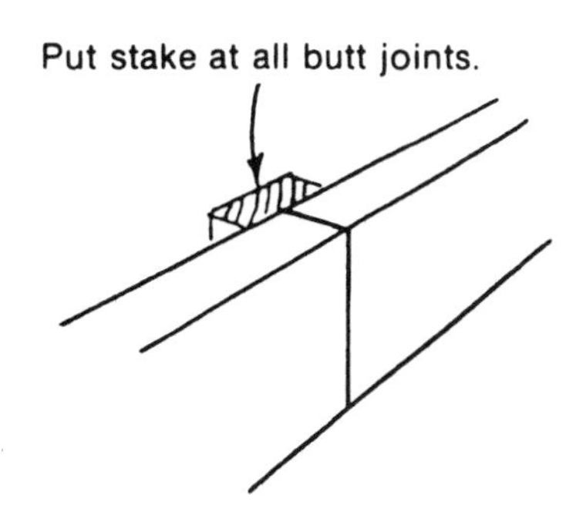

After nailing, cut off stake for easier finishing.

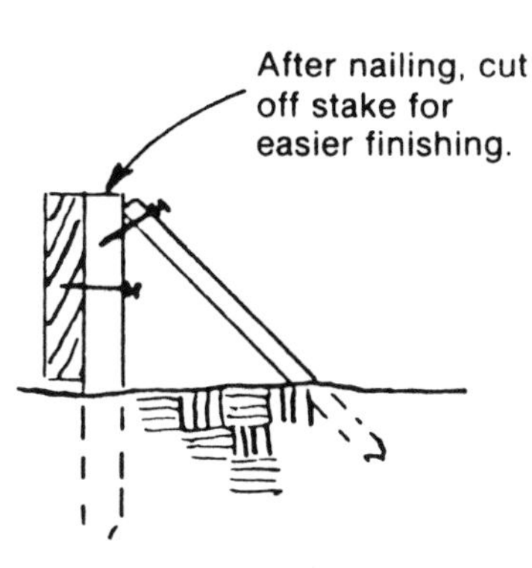

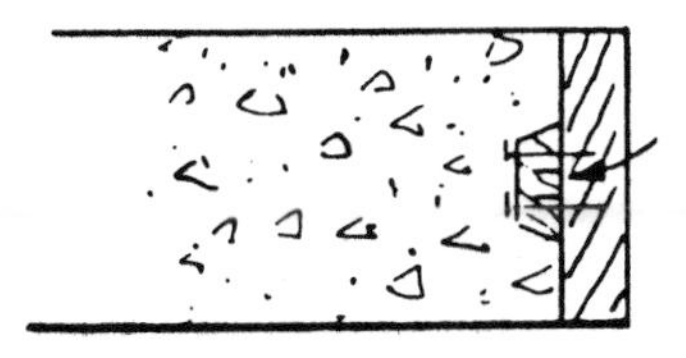

When forming a concrete construction joint, saw-cuts here will make stripping easier. Metal keyways and premolded keyways (left in slab permanently) are also available.

For slabs 4 inches thick (or thinner) steel dowels should be used in place of keyways.

Treated wood (redwood, cypress, or cedar) divider can be used for a control joint.

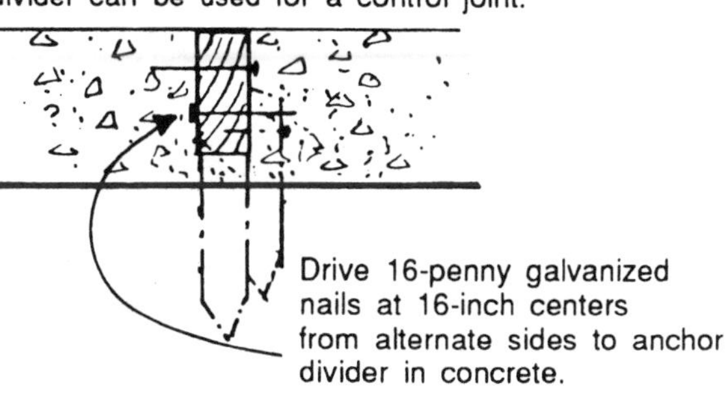

Drive 16-penny galvanized nails at 16-inch centers from alternate sides to anchor divider in concrete.

Forming a Curved Slab or Curb

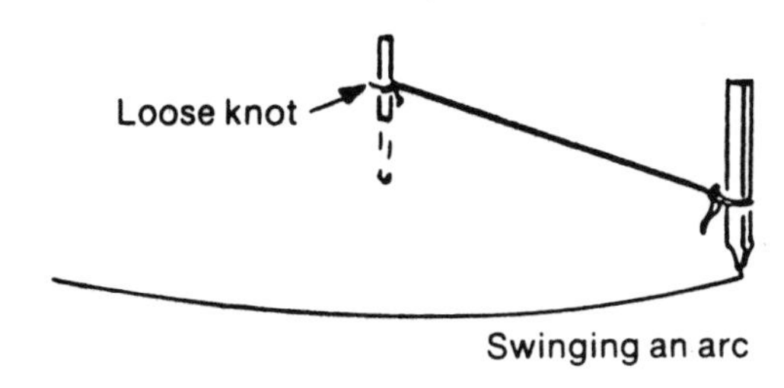

Swinging an arc

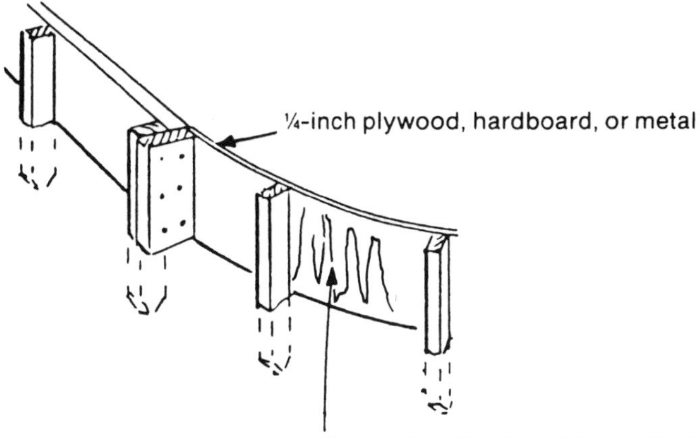

Plywood grain should be vertical for short radius curves.

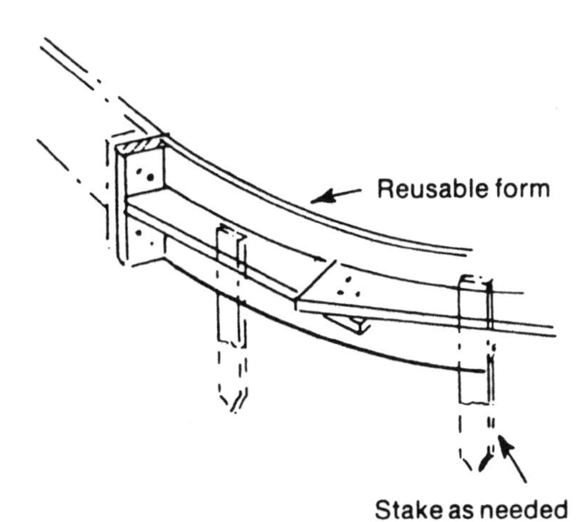

Stake as needed

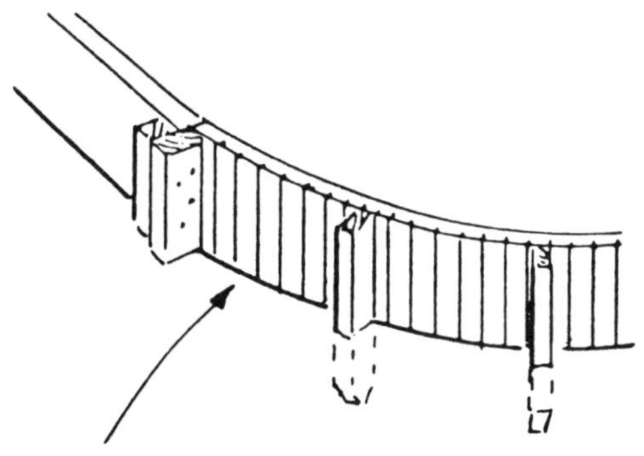

For easy bending, ¾-inch material can be kerfed. Sometimes the bottom is kerfed on uneven ground. Soaking the wood overnight makes bending easier.

Forming Walls

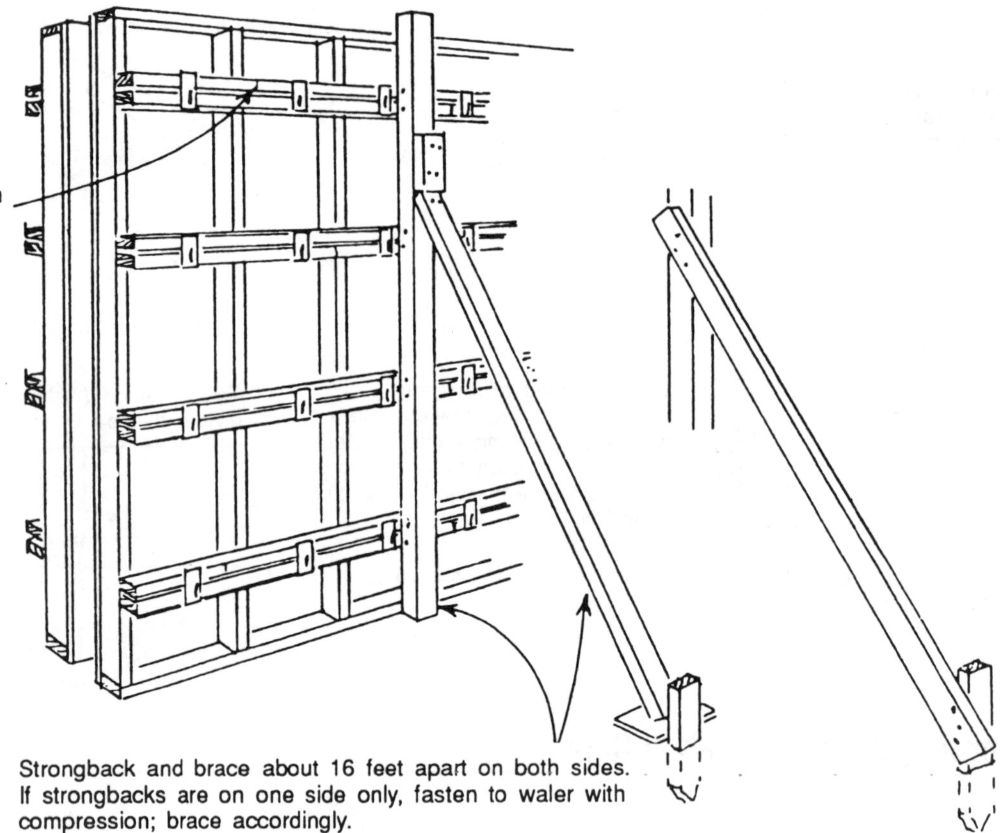

(Dimensions given are for illustrative purposes only.) Dimensions required for bracing may be affected by the height of the wall, the rate of the casting, and other factors.

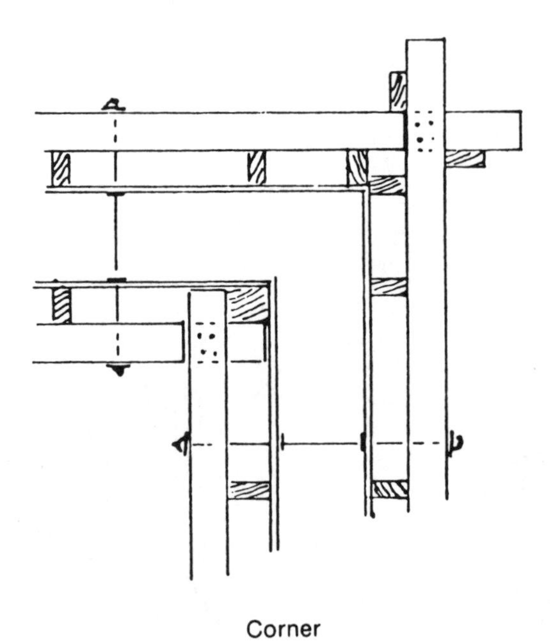

Corner

Reusable Job-Built Plywood Form Panels

Some builders make standard plywood forms to be used on many jobs. These plywood panels are gang-drilled for ties. Careful use and cleaning prolongs the life of plywood form panels.

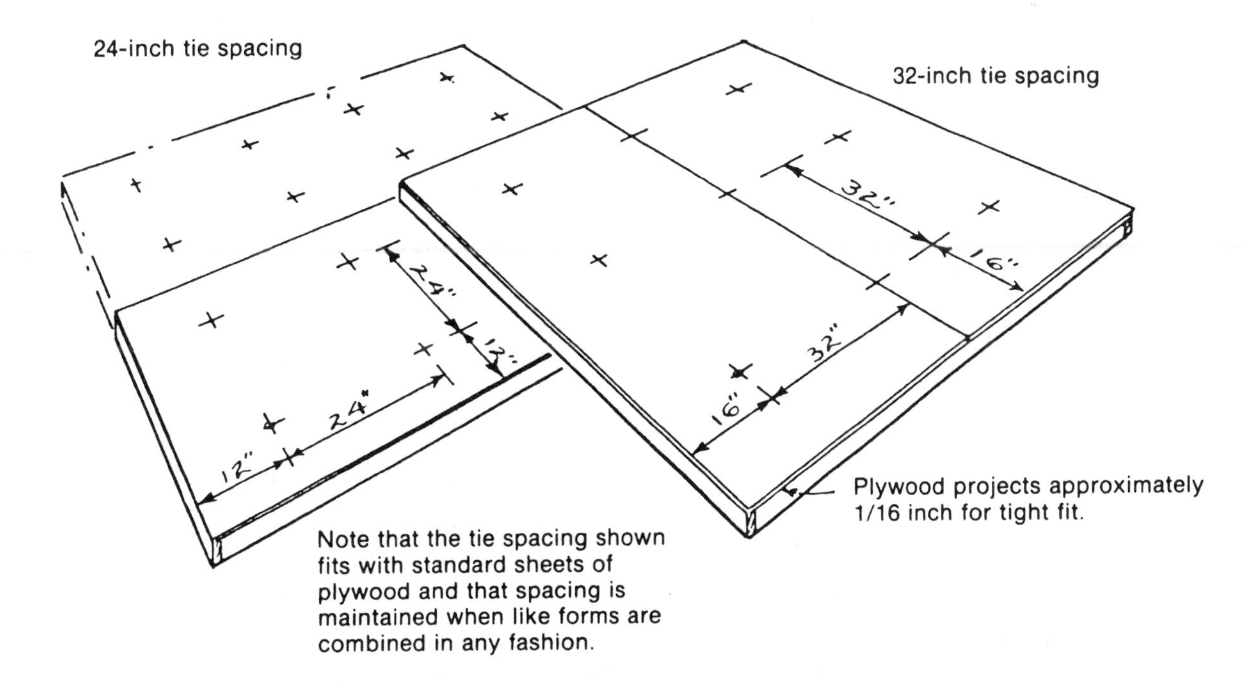

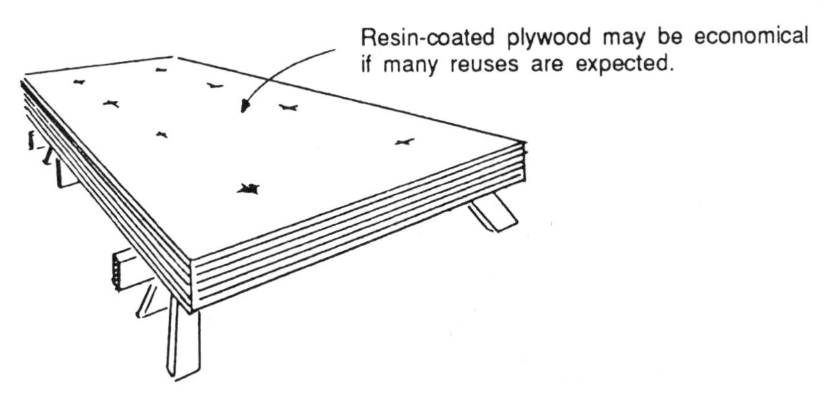

Reusable Prebuilt Plywood and Steel or Aluminum Frame Forms

Labor requirements for erecting and stripping reusable prebuilt plywood and steel or aluminum frame forms are only about ⅙ to ⅛ of that needed with typical job-built forms.

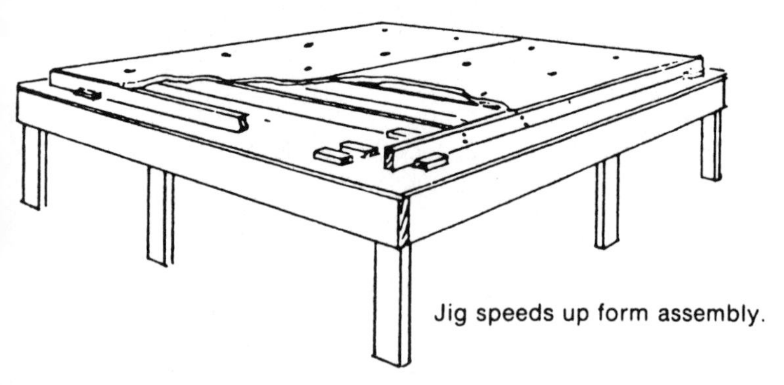

Jig speeds up form assembly.

One forming system uses HDO (high-density overlay resin) plywood panels. These panels are manufactured in 2x8-foot sections and are 1 ⅛ inch thick. The system features reusable hardware with spreadable cleats and snap ties. Walers and strongbacks are not needed in most cases. Special techniques are involved in securing the corners. Manufacturers guarantee a number of reuses for the panel pieces. The cost per panel is higher than for conventional forming plywood; however, the ease of installation, stripping, cleaning, and reusability result in a low cost per house.

Some panels are cut into various lengths to use as fillers for spaces shorter than the 8-foot length. Several other techniques in the use of this system save labor and forming material. The panels can be easily handled and carried by one worker.

Many forming systems are patented, and detailed information about specific systems can be obtained from the manufacturers.

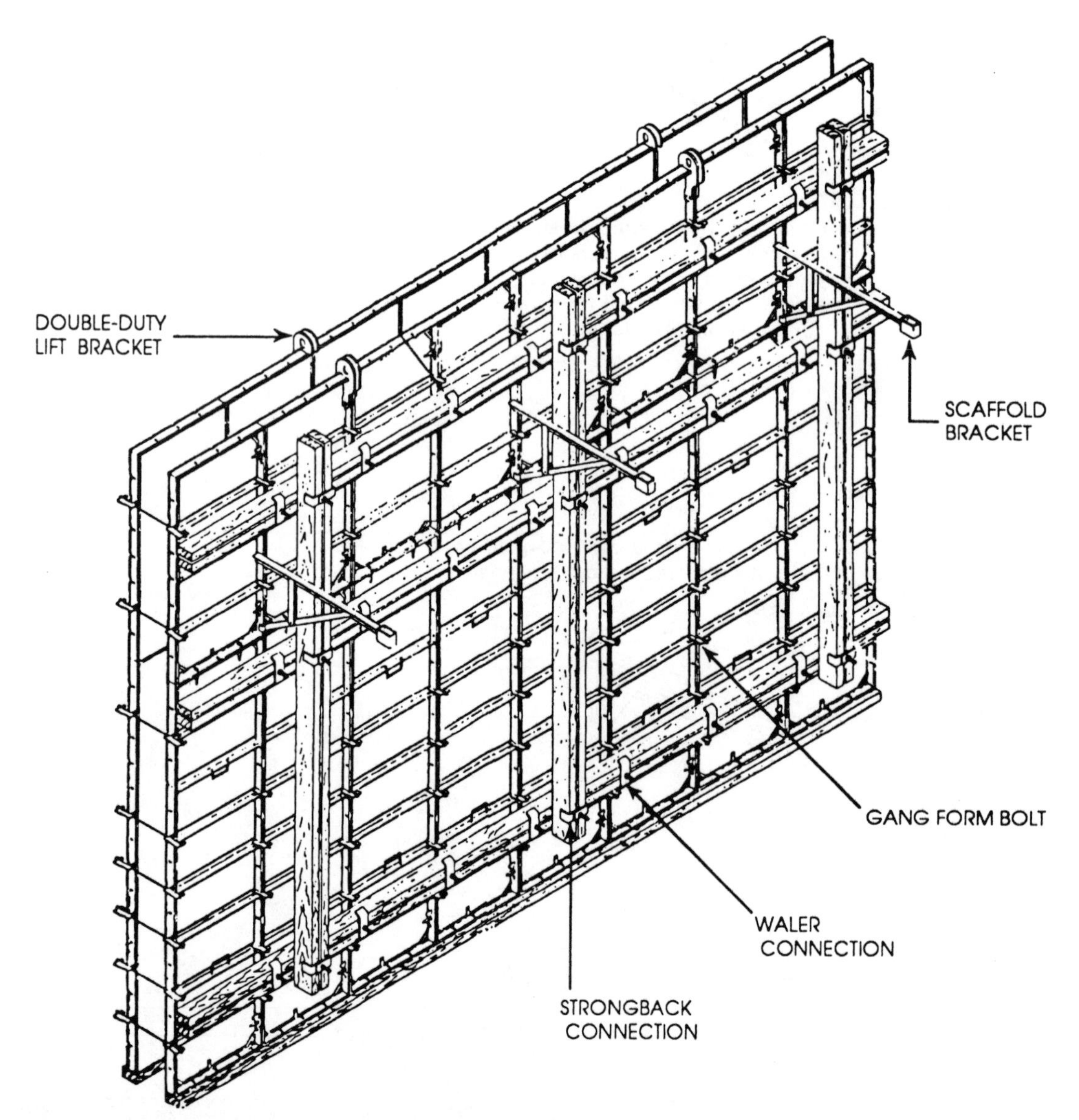

Reusable prebuilt plywood and steel or aluminum frame forms.

Polystyrene Forming Systems

Polystyrene (foam) forms provide a fairly new way of building walls. In one method, after a conventional footing is placed, polystyrene hollow blocks are assembled as interlocking units 16 to 48 inches high. The blocks are braced and wired down; vertical and horizontal rebars are inserted in the hollow units; and then concrete with 4- to 5-inch slump is placed into the blocks. When the concrete sets, the complete grid of beams and columns forms a structural wall that is fully insulated. Inside and outside finish can be applied to the polystyrene. Among the advantages of a foam forming system are—

- Forms become part of the wall, minimizing waste and eliminating stripping.
- Foam forms may provide insulation for added energy efficiency.
- Foam forms are lightweight and easy to cut and place.
- Stay-in-place forms protect the concrete during curing, promoting good curing and strength.
- Foam forms save lumber resources.
- The finished wall has good sound control.
- Foam forms can save construction time and costs.
- Using foam forms requires a minimal level of construction skill.

High-density overlay resin plywood panels used to form a footing. Note overlaps used to prevent cutting of the panels.

Forming Door and Window Openings (Blockouts)

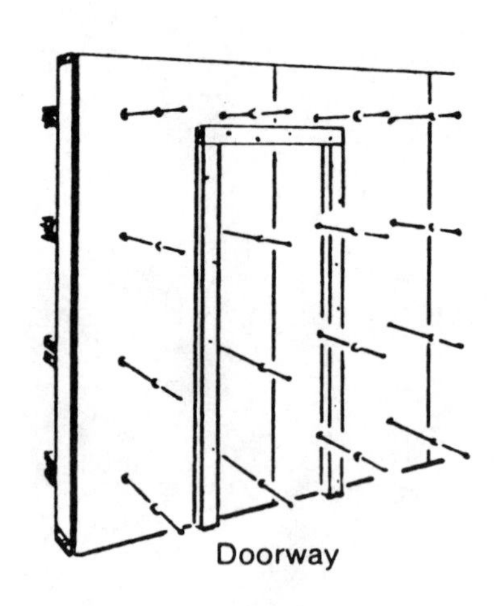

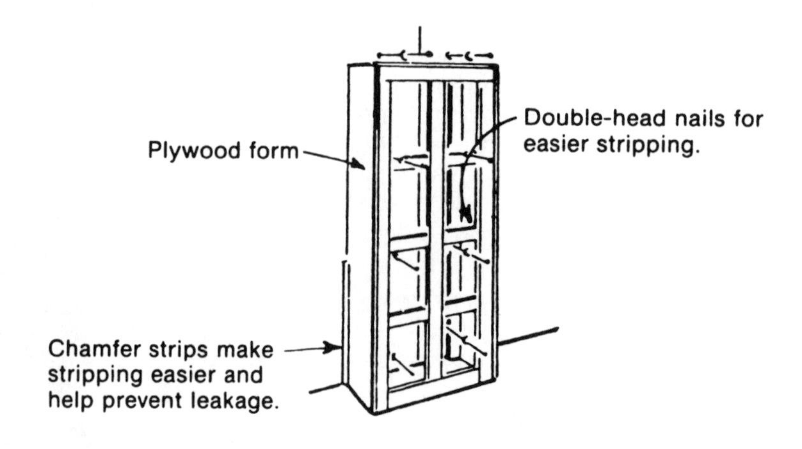

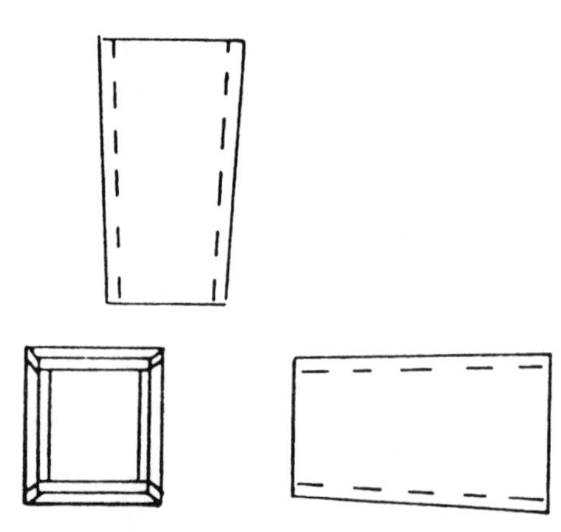

Form for small openings tapered for easy removal.

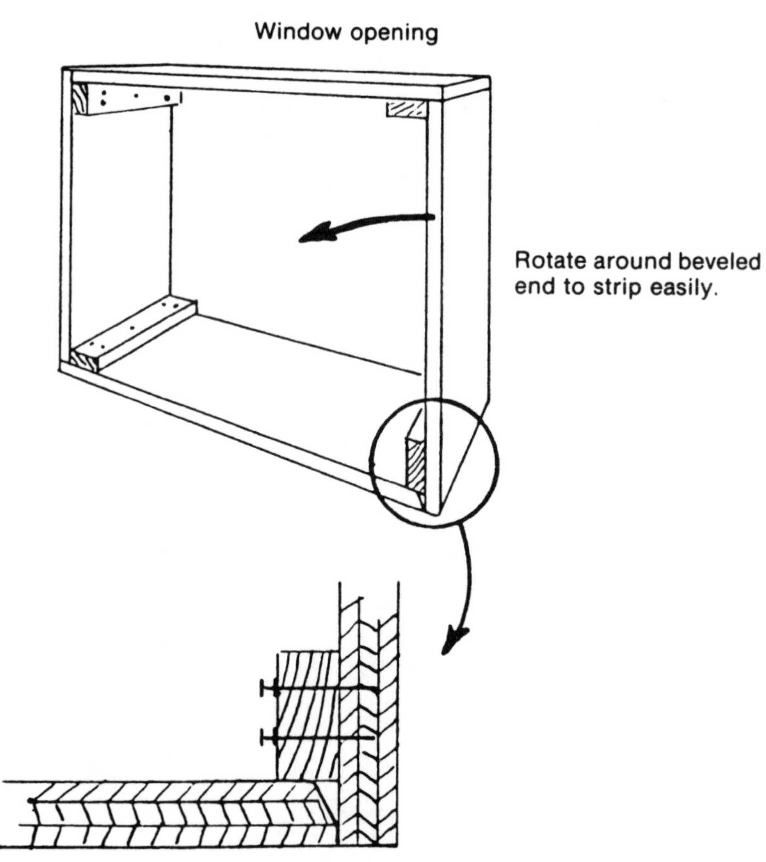

Forming a Single Wall

Bracing can be reduced by installing treated wood forms on the blind side and leaving the form in place. Form ties contain the pressure of the wet concrete mix. Snap ties are combination form spacers/tie-bars that have plastic cones at the form face so they can be broken off beneath the surface and patched.

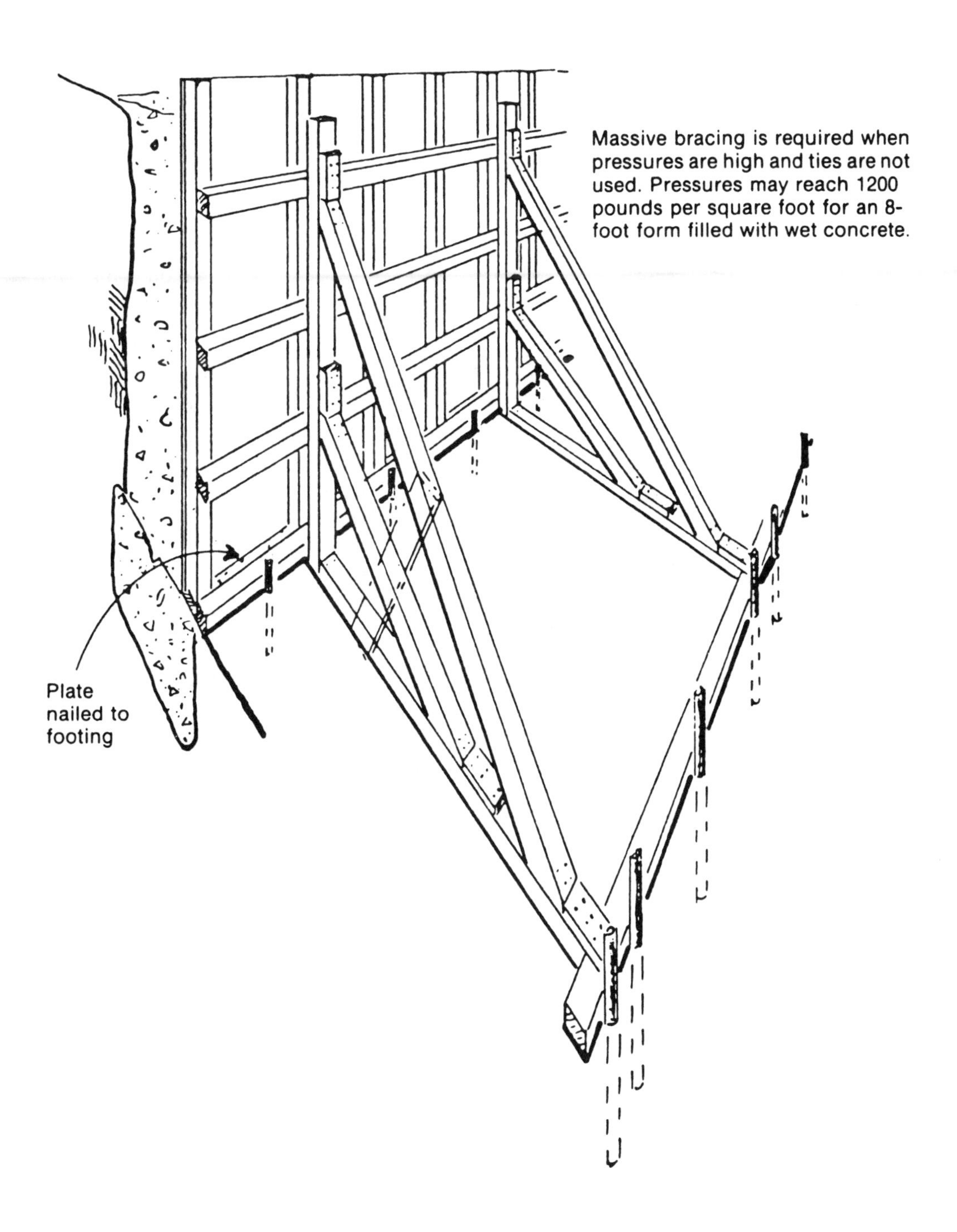

Forming Stairs

Typical Low Exterior Stairs

Stairs and Safety
Higher steps might have a 6-inch riser and 12-inch tread including tread nosing. For safety there should be no variation in the heights of risers or the widths of treads in a set of steps.

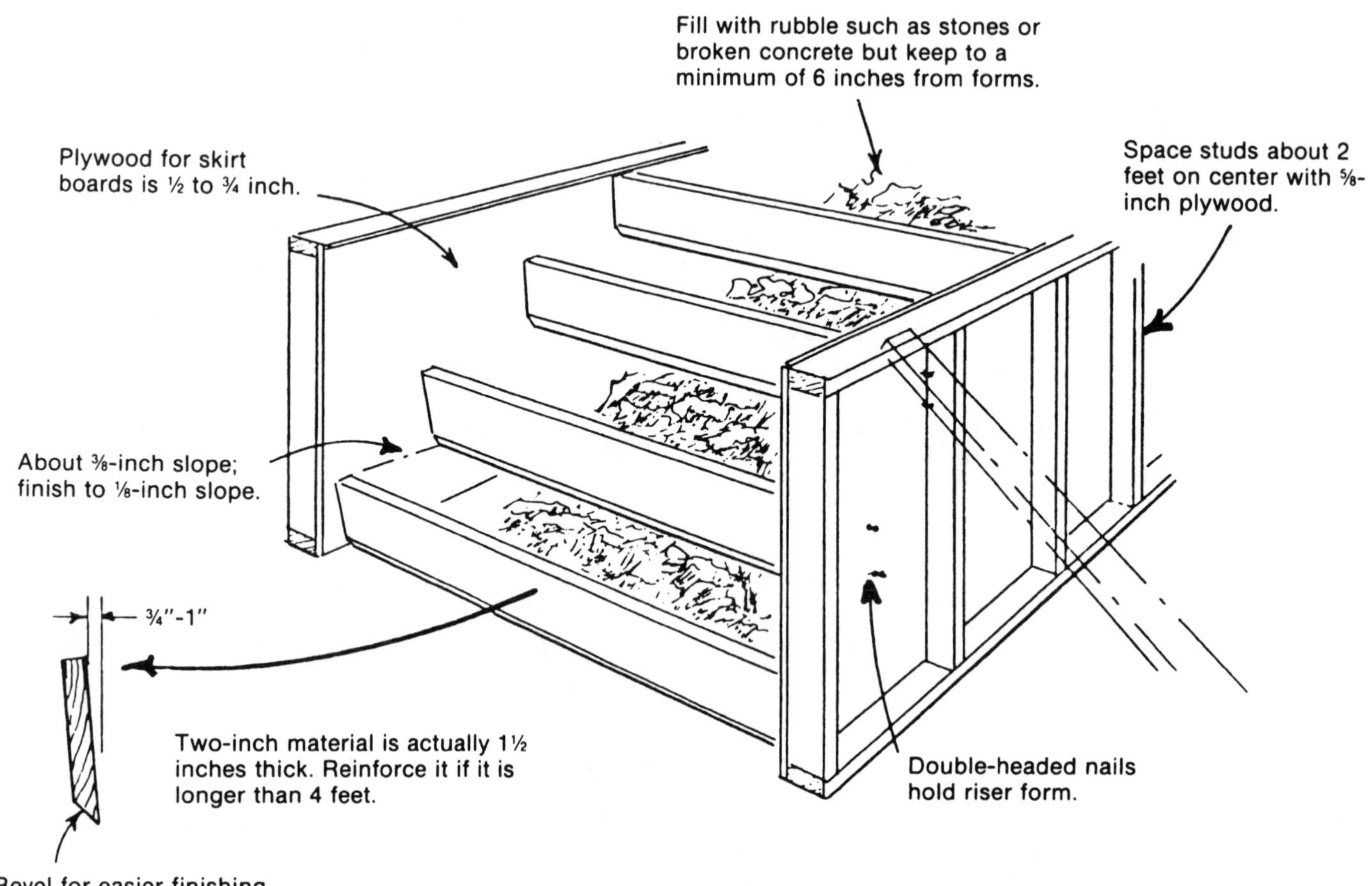

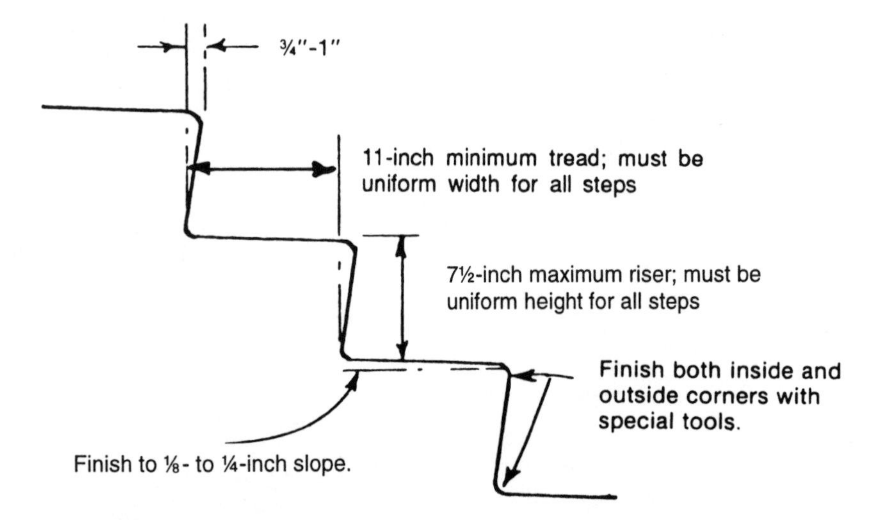

Riser Bracing for Wide Stairs

Bracing should be used with riser forms when the riser form material is nominally 2 inches thick (1 ½ inches actually) and spans more than 4 feet.

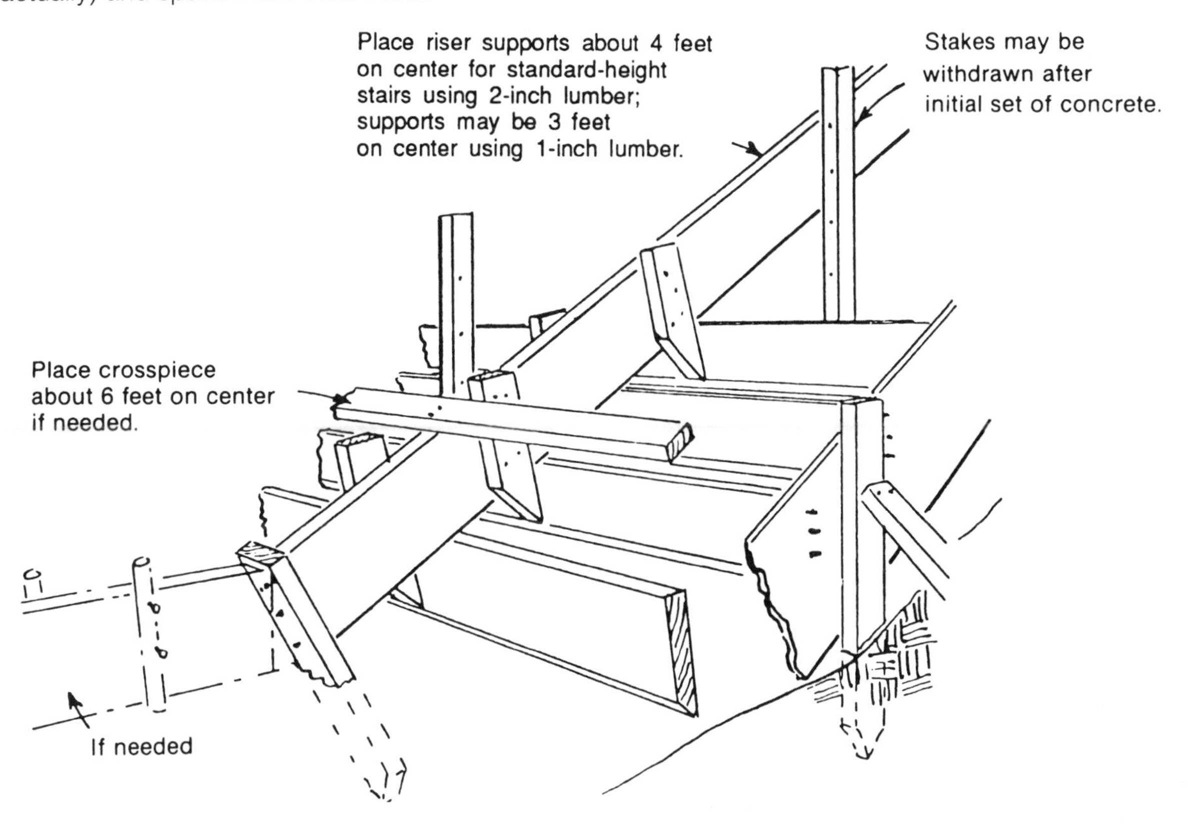

Fastening Riser Form to Concrete or Masonry Wall

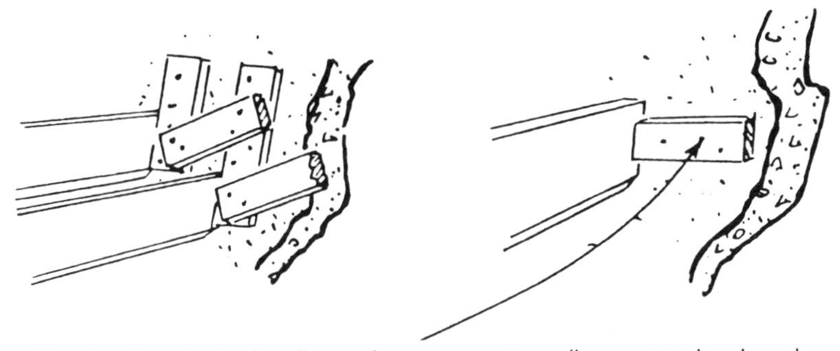

Use hardened steel nails such as concrete nails or case-hardened cut nails to fasten blocks to concrete.

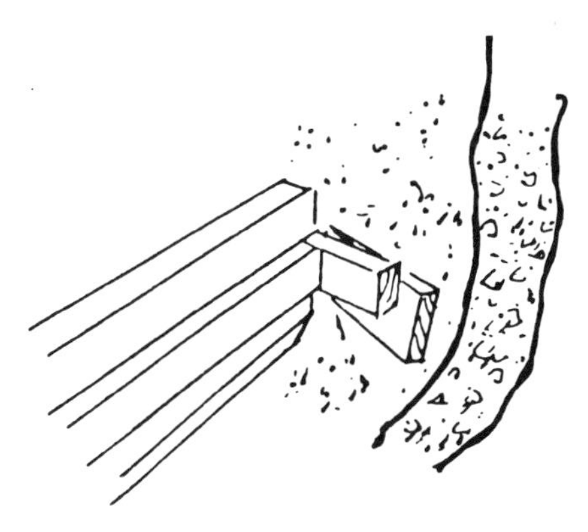

Alternate: Wedge the block in place.

Stair Support

Stairs for both new and old construction should rest on firm supports—undisturbed soil or compacted fill at least 6 inches below the frostline to prevent settling, cracking, and pulling away from the building.

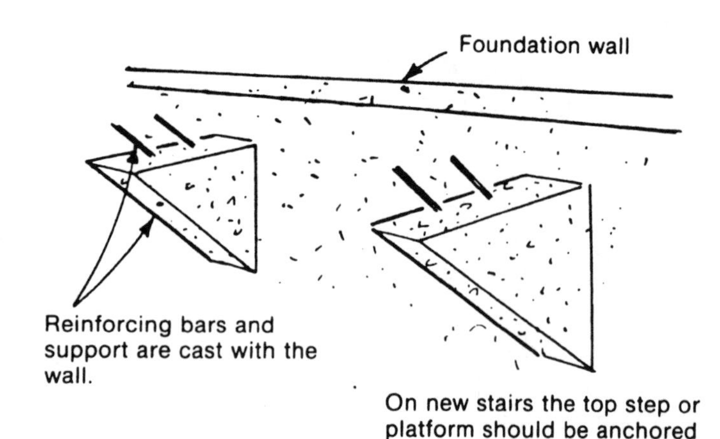

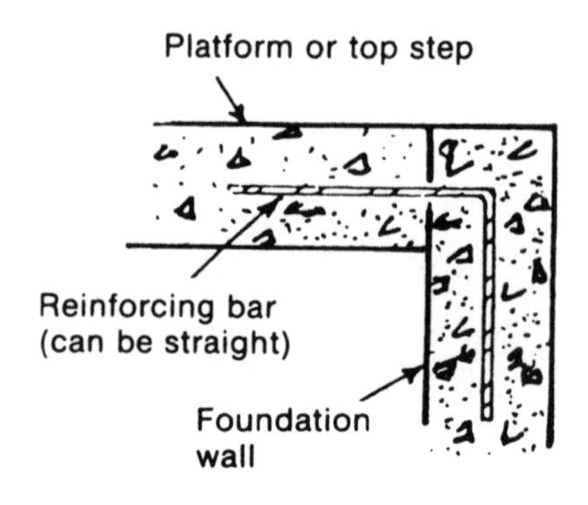

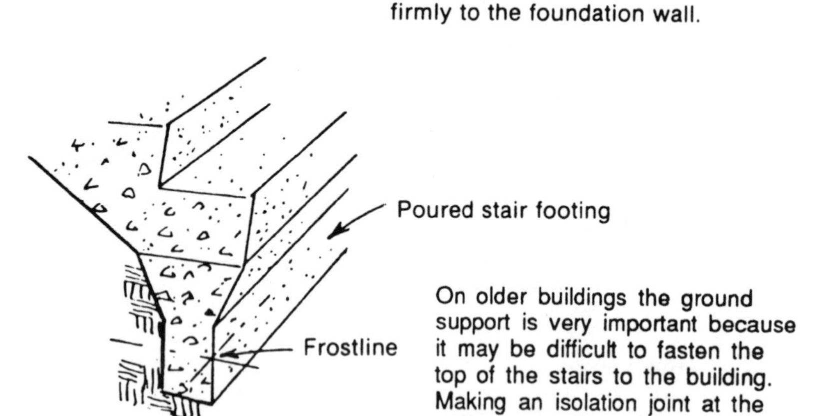

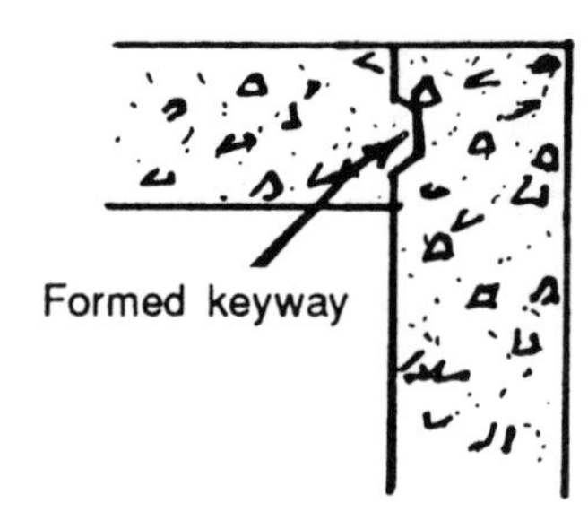

Alternate ways to fasten to new construction

Open Stairs

Stairs supported at the top and bottom must be designed structurally for thickness with reinforcing steel bars according to span and load.

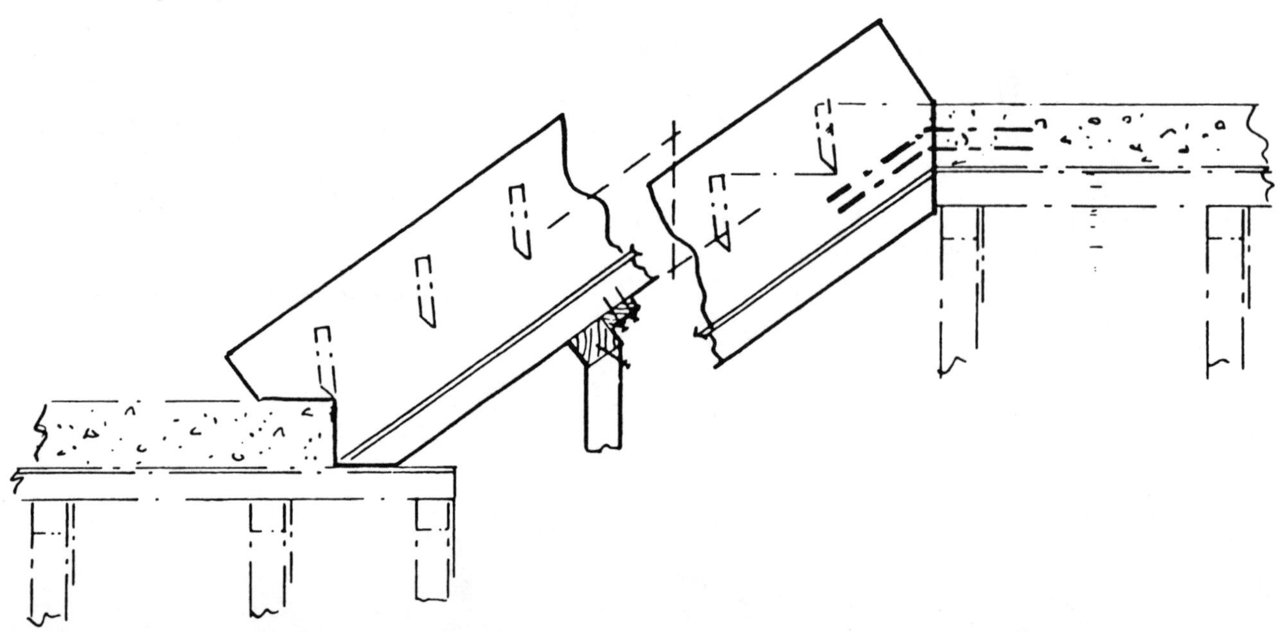

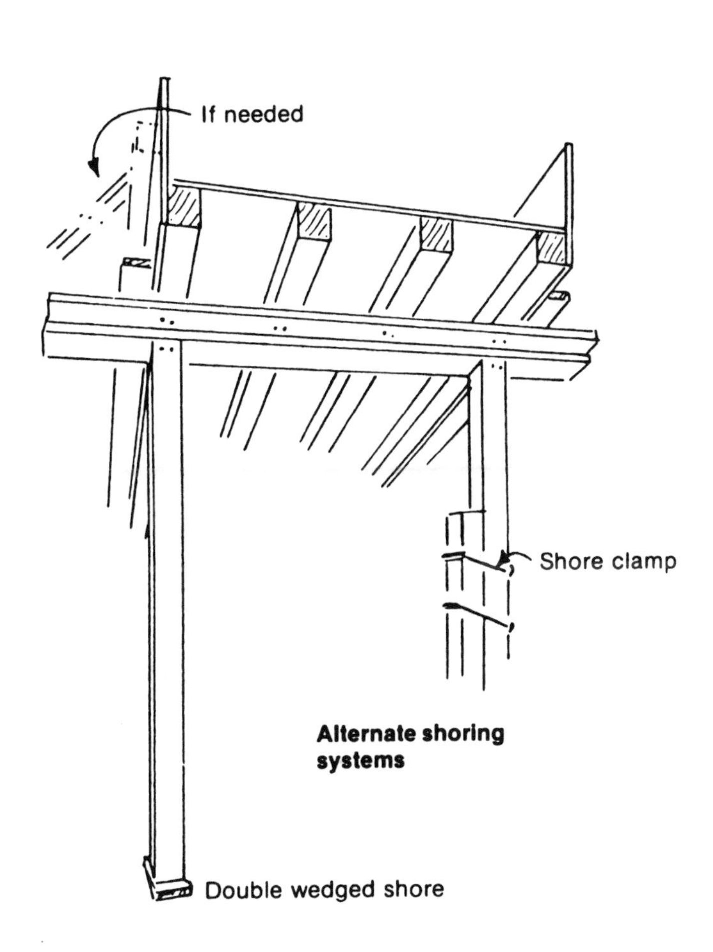

Alternate shoring systems

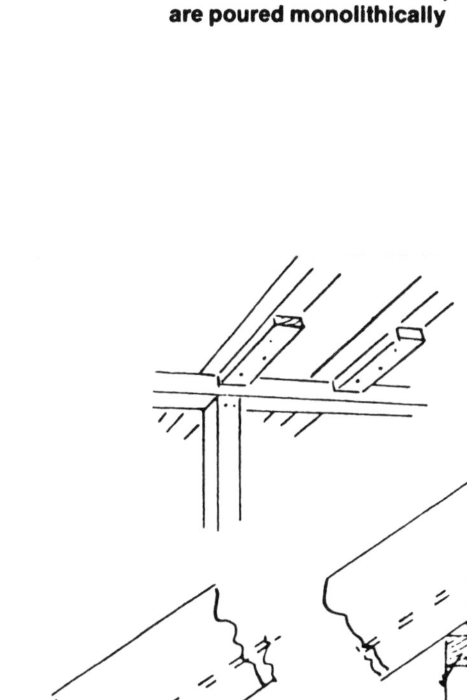

Alternate: when stair and platform are poured monolithically

Alternate kicker systems

Inspection

Most municipalities require an inspection before concrete is placed. Checklists are sometimes used to be sure that the building code is met. Items on these checklists might include—

- Water and debris are removed from forms.
- Footings, foundations, and slabs conform to the plans.
- Subgrade is properly shaped, compacted, and not frozen.
- Materials conform to applicable codes and standards.
- Forms are true, tight, and properly braced.
- Under-slab ducts and plumbing are as planned.
- Reinforcements—rebars or wire mesh—are properly located and tied.
- Treatment for termite protection has been done if needed.

Preparations have been made as needed to compensate for hot or cold weather.

Summary of Forming Concrete

Forms basically act as a mold, holding the wet concrete in place until it has built up sufficient strength to retain its position and shape on its own. A variety of forming materials and techniques are available for use by residential builders. When planning your job, keep in mind the following—

- Consider reusable forms (remembering to factor in time to set up and disassemble them).
- Look for new materials and techniques that can save time by cutting steps. Contact other builders who have tried new methods and ask them about their experiences to find the best "fit" for your jobs.
- Be sure that form materials are thick enough and braced to sustain the pressure from the concrete.
- Stairs should rest on a firm bed of undisturbed soil or compacted fill at least 6 inches below the frostline.
- Before placing concrete, check to be sure reinforcements are properly located and tied.

Chapter 4

The Concrete

The most convenient way to obtain concrete is to order it from a local ready mix concrete supplier. The ingredients for ready mixed concrete are accurately measured, often in automated plants with computer equipment, and then mixed in a truck enroute to the jobsite.

Ordering Ready Mixed Concrete

Ready mix concrete producers will generally assist in the selection of appropriate mixes or in developing mixes to fill a builder's specific needs. Mixes must be proportioned to provide the strength required for the type and location of the construction, as shown in the following guidelines. Local practices in ordering ready mixed concrete vary from area to area, but usually ordering is done by specifying the size of the job and the desired strength and performance of the concrete.

Builders should advise the ready mix concrete producer of the intended use for the concrete (such as a footing, sidewalk, or basement slab). Quite often suppliers have standard mixes for specific applications (such as a sidewalk mix, a footing mix, or a curb and gutter mix). By specifying its intended use, builders can help to ensure that the proper concrete is delivered.

When the builder specifies a particular strength level, the ready mix concrete producer is responsible for proportioning and delivering a mixture that will yield the desired strength. For engineered construction, properly conducted strength tests are required.

Concrete also can be ordered by prescription. Using this method, the buyer specifies the weight of portland cement per cubic yard of concrete, the maximum

Guidelines for Selecting Concrete Strength

(Recommended by ACI Committee 332)

Specified Compressive Strength (f_c') at 28 days, psi*

Type or location of concrete construction	Regional weathering areas**			Desired Slump (in.)
	Negligible	Moderate (air)	Severe (air)	
Basement walls and foundations not exposed to weather	2,500	2,500	2,500	6±1
Basement slabs and interior slabs on grade	2,500	2,500	2,500	5±1
Basement walls, foundations, exterior walls, and other concrete works exposed to weather	2,500	3,000 (5-7%)	3,000 (5-7%)	6±1
Driveways, curbs, walks, patios, porches, and unheated garage floors exposed to weather	2,500	3,000 (5-7%)	4,000 (5-7%)	5±1

*Specifications ASTM C 94 permit 10 percent of the strength tests to fall below f_c', *but no tests can be more than 500 psi below.*

**See map on page 37.

When selecting the type of concrete, builders should consider the durability and finishability needed for the job conditions.

amount of mixing water, admixtures required, and possibly their dosage rates. If concrete is ordered this way, the buyer accepts sole responsibility for the quality and performance of the concrete although the ready mix supplier remains responsible for accurately batching and adequately mixing the ingredients. Nowadays, concrete for residential construction is rarely ordered by prescription.

The builder must also tell the supplier where and when the concrete is to be delivered, and the amount required. An order should specify about 5 percent more concrete than the computed volume of the forms to allow for spillage, form movement, and consolidation. Accurate calculations and form building are essential for ordering correct quantities of concrete. Good record keeping will allow a builder to compare estimated quantities with the actual amount of material used, allowing an adjustment to be made to the 5 percent increase in the quantity ordered.

Many builders have traditionally ordered by "sacks" or "bags" per yard (for example, a "six sacks mix"). This method, while still used in some areas, is becoming less accurate as new additives are included in cement and concrete mixes. Suppliers suggest that rather than rely on this general method, builders discuss the parameters of the work with their supplier to determine the order.

Concrete is a perishable product. To do a good concrete job, builders and ready mix concrete suppliers must work together to minimize delays. Before the concrete arrives, a builder should check the formwork to ensure that it is accurately set and adequately braced to withstand the pressure of fresh concrete. Make sure that an experienced crew is properly equipped to handle, place, finish, and cure the concrete, and that the ready mix concrete trucks have easy access to all locations for placement. Because they are so heavy, ready mix concrete trucks cannot drive over curbs, sidewalks, or driveways without causing serious damage. If the builder insists that a concrete truck pass over these constructions, a signed liability waiver will usually be required.

When the concrete arrives at a job, the builder should be given a delivery ticket with the following information—

- Ready mix company name, batch plant, and load number
- Class or designation of concrete conforming to the job specification (example: 3,000 psi, air entrained)
- Amount of concrete, in cubic yards
- Location of this batch on the jobsite (important when multiple placements are scheduled at the same site)
- Time of loading (generally ready mixed concrete should be discharged from the truck within 1 ½ hours of loading)

Upon arrival at the jobsite, one addition of water may be permitted to bring the load to the specified slump, but only if the addition does not make the mix exceed the specified water-to-cement ratio. Most ready mix concrete producers require builders to sign for water added at the jobsite because the strength of the concrete may be affected by the added water. The water should be thoroughly mixed into the load and the concrete discharged as quickly as possible. Additional water added as a load is discharged will reduce the quality of the concrete.

Jobsite mixing of concrete is sometimes necessary because of remote location, limited access by the ready mix truck, or unusual site problems.

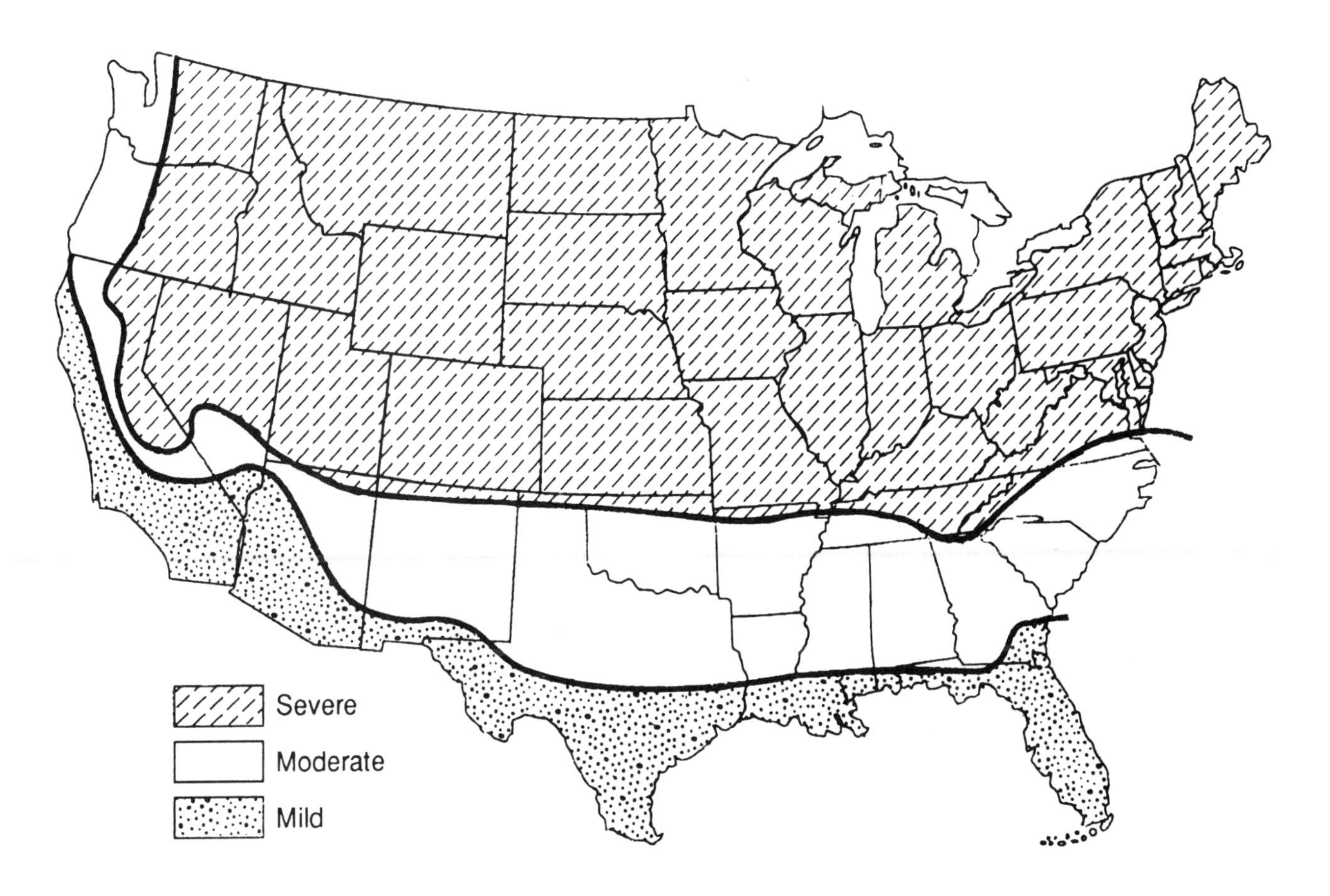

Weathering Regions for Residential Concrete
The weathering regions map provides a guideline to the location of severe, moderate, and mild winter weathering areas throughout the United States (Alaska and Hawaii are classified as severe and mild, respectively). The map cannot be precise. This is especially true in mountainous areas where conditions change dramatically within very short distances. It is intended to classify as severe any area in which weathering conditions may cause deicing salt to be used, either by individuals or for street or highway maintenance. These conditions are significant snowfall combined with extended periods during which there is little or no natural thawing. If there is any doubt about which of two regions is applicable, the more severe exposure should be selected.

The three exposures are—

1. Severe: Outdoor exposure in a cold climate where concrete may be exposed to the use of deicing salts or where there may be a continuous presence of moisture during frequent cycles of freezing and thawing. Exposure to severe weather can damage pavements, driveways, walks, curbs, steps, porches, and slabs in unheated garages. Destructive action from deicing salts may occur either from direct application or from being carried onto an unsalted area from a salted area, such as on the undercarriage of a car traveling on a salted street but parked on an unsalted driveway or garage slab.
2. Moderate: Outdoor exposure in a cold climate where concrete will not be exposed to the application of deicing salts but will occasionally be exposed to freezing and thawing.
3. Mild: Any exposure where freezing and thawing in the presence of moisture is rare or totally absent.

Placing the Concrete

Place concrete as close as possible to its final position. Do not try to move it horizontally over long distances—especially if it is high slump. Working the concrete in this manner can result in segregation of the aggregate from the paste; this thin watery cement paste may then be worked to the surface, causing later crazing, dusting, or scaling.

Place new concrete against the face of any previously placed concrete, rather than in separate mounds. Separate mounds of concrete can partially set up before they are joined, which may cause a poor bond.

Start placing concrete for walls or slabs at one end and work across toward the other end. For walls, work in lifts of 1 to 2 feet deep. For slabs, work in strips of 4 to 6 feet wide. Avoid collecting water at ends, in corners, and along form faces.

Use a baffle on slopes to avoid separation and accumulation of aggregate at the bottom of the slope. Place concrete on the lower part of the slope first and move upward.

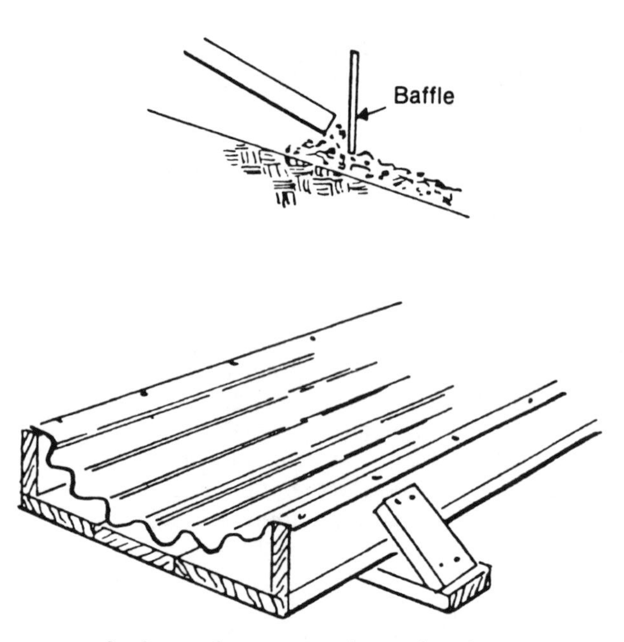

A piece of corrugated metal makes a good chute. So does a plywood bottom with 2x8 sides.

Do not drop the concrete more than 4 feet to its final location to prevent aggregate segregation. If a longer drop is required in forms such as basement walls, use a tremie or chute. Do not drop the concrete so that it hits the sides of the forms or through congested reinforcement. Serious segregation of the aggregate may occur. Careful placement using chutes or "elephant trunks" are recommended. An elephant trunk is a 4- to 6-inch-diameter tube of rubber or canvas that keeps the concrete from segregating because it drops the mix from a hopper to the final location. This equipment often can be supplied by the cement finisher or concrete subcontractor.

Remember to use only accepted concrete-working tools such as short-handled square-end shovels or concrete rakes. Do not use yard rakes; they may increase segregation. A mason's pointed hand trowel is useful for some concrete work; however, it should not be used to finish concrete.

These are the basic tools of the trade for hand-finishing concrete. At the top is a straightedge, and just below, a darby. From the left is a finish trowel, a jointer (or groover), an edger, and a magnesium float.

Power finisher-float.

Mechanical vibration tools (vibrators) can be very useful to remove air bubbles from the concrete, especially on walls and other deep placements of concrete. Large air bubbles and poor consolidation can create troublesome voids—called "honeycomb"—in the finished concrete. Skillful vibration of each 12- to 24-inch lift of concrete in a wall will help consolidate the lower lift with the upper lift and prevent honeycomb.

Vibration of concrete provides good consolidation and improves the bond of concrete to reinforcement. In turn, good consolidation and bonding reduces the probability of bugholes, rock pockets, and other problems. When using a vibrator, the operator should watch the affected area while working the vibrator head in an up-and-down motion, going slowly and as deeply as possible. Always pull the vibrator out of the concrete slowly to prevent formation of a hole or bubble in the concrete. Use of a vibrator is not recommended for exposed-aggregate surfaces since the action of the machine can cause bare spots (also called sand spots).

The jitterbug—a tool used to depress the coarse aggregate down from the top surface of a concrete slab—leaves a thin coat of sand, cement, and water on the surface and makes finishing easier. However, it also leaves a weaker concrete surface that is subject to spalling and scaling. Therefore, the use of a jitterbug is strongly discouraged. Any greater ease in finishing is likely to be offset by the complaints of the homeowner when deterioration of the concrete slab begins in as little as 2 to 3 years.

Make it a practice to strike off the top surface as soon as the concrete is placed to avoid disturbing the concrete later on.

Placing Concrete for Steps

To place concrete in stair forms, begin at the bottom step and work upward. Carefully spade or vibrate the mix, especially next to form faces. Each tread should be struck off level as it is filled. Forms should be tapped lightly to release air bubbles.

Concrete Safety

Concrete is generally a safe building material. However, concrete is abrasive, and wet concrete is very caustic and can cause skin burns or cement dermatitis. A few simple precautions can protect you and workers on your site from injury or skin irritation. Insist that all workers handling concrete protect their eyes by wearing safety glasses with side-shields or goggles. Workers also should protect their skin by avoiding prolonged contact with the fresh concrete mix and by wearing long-sleeved shirts, full-length pants, rubber boots, rubber gloves, and kneepads.

Should skin irritation occur, wash the affected area thoroughly with soap and water and then apply vinegar, which can help neutralize the caustic content of cement. For a severe burn, or if the condition persists, consult a physician familiar with cement burns.

If the eyes are affected, wash them with clean water and consult a doctor as soon as possible. Do not use any other treatment without consulting a doctor. Do not rub the eyes.

While serious burns from concrete are rare, the best way to prevent injuries is to be sure that all workers handling concrete understand the potential hazards and follow appropriate precautions.

Finishing the Concrete

Complete striking off and bullfloating before bleed water appears, and do not begin finish floating and troweling until most bleed water evaporates or is removed. If the concrete is worked while bleed water is present, a water-rich mix that is sand heavy is created at the worst place—the surface. Dusting, scaling, and crazing are almost certain to result. For the same reason, do not add water to the surface for ease of finishing.

Sometimes concrete begins to set up while bleed water is still present. In that case finishing must begin, and one of the following steps can be taken—

- Go out on the concrete on kneeboards and sweep off the water with a long trowel, darby, hairbroom, or squeegee.
- With one person at each end, drag the surface with a rubber hose.
- Lay damp burlap on the concrete and sprinkle dry cement on the burlap. Later, throw away the cement; the burlap can be washed and reused.

Do not add cement directly to the surface to take up the water; it is likely to cause the surface to crack, craze, or dust after it dries. If air-entrained concrete with a low water-cement ratio is used, excess bleeding should not be a problem.

Striking Off and Bullfloating

Concrete flatwork should be struck off, or "screeded," as it is placed, and bullfloated as soon as it is in place. Complete bullfloating before bleed water appears. Take care that you do not let any one finishing operation get too far ahead of the others.

To avoid segregation, do not vibrate the wet concrete mix to level it. If the finishers have trouble working the concrete into place, water reducers or high-range water reducers should be considered for the next slab.

If possible, hold the bullfloat perpendicular to the strike-off direction and bullfloat a second time perpendicular to the first. These measures help to avoid dishing the surface. If dishing occurs, the surface may need to be leveled with a darby, which is a long, hand-held strike-off tool. Striking off with a full-length screed also will help prevent dishing.

Wood bullfloats tend to be better for opening up the concrete and letting bleed water out—for example, on concrete that is not air-entrained in cool or damp weather. Wood bullfloats are also better for shaving off and leveling the surface. Magnesium or aluminum bullfloats are easier to use on air-entrained concrete, and they help to avoid tearing the surface. To minimize segregation of the concrete, work the surface no more than necessary.

Edging

Edging is easier if the concrete is cut away from the form with a trowel beforehand. A good time to do this is right after bullfloating and before bleed water appears.

Wait until the concrete is ready for finishing before starting the edging. A good test for readiness is when the weight of a person standing on the concrete does not make footprints deeper than ¼ inch. Bleed water should be gone. Use a wide-flanged edger for the first pass to prevent making a deep mark and to avoid sealing off the surface too much.

In very hot weather when the concrete is setting fast, the first pass with the edger can be run at the same time the concrete is cut away from the form with the trowel. This step should be taken soon after bullfloating.

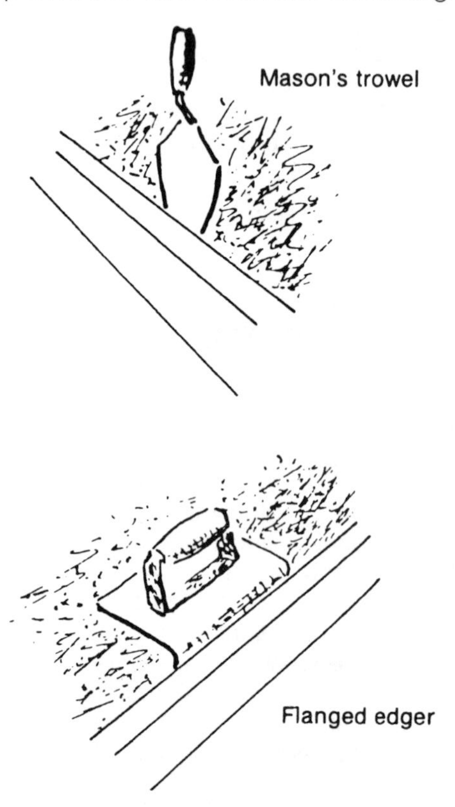

Making Control Joints

For guidelines on laying out control joints, see Chapter 5. Control joints should be cut to ¼ of the slab depth but not less than 1 inch deep. Finishers often do not go this deep since it requires more work. But when they do not, the cracks go elsewhere. If the cut is too deep, aggregate interlock is lost and the concrete sections may then slip past each other vertically.

Remember that wire mesh should not be lapped at control joints.

The joints can be hand-tooled, sawed, or formed with pressure-treated lumber or asphalt-impregnated strips that are left in place. Remember to allow control joints time to widen from drying shrinkage of the concrete before sealing.

Control joints must be at least one-fourth of the slab depth to be effective.

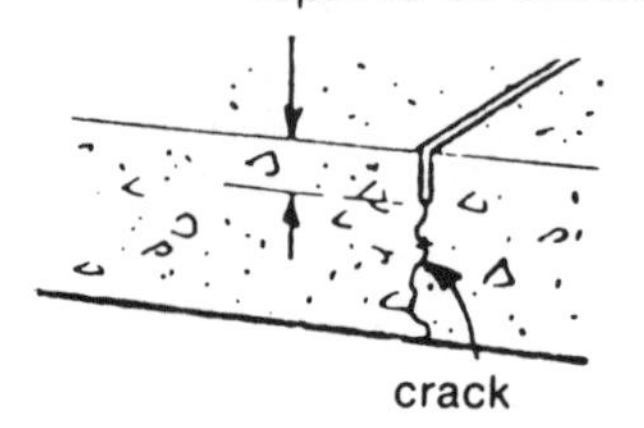

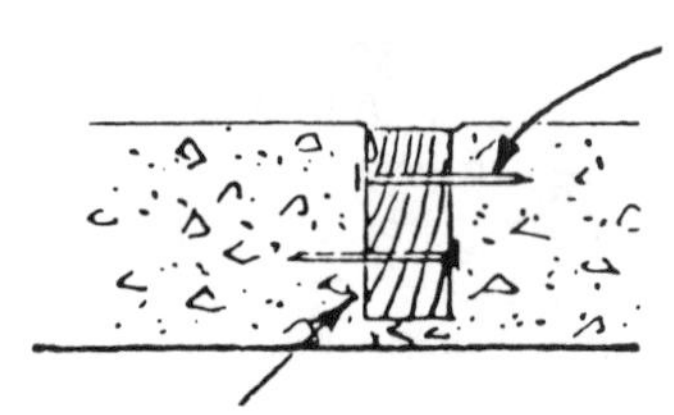

Galvanized nails at 16-inch centers driven from alternate sides

Control joint of nondecaying lumber

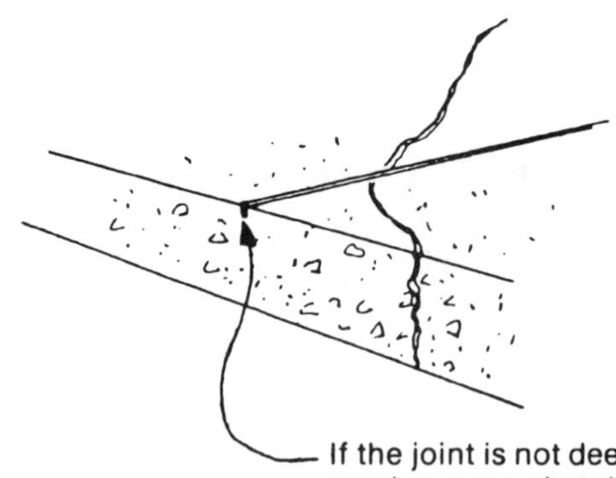

If the joint is not deep enough, the crack may go elsewhere. If it is too deep, the aggregate will not interlock and displacement can occur.

Sawing the Joint

Concrete saws are readily available for rent and are increasingly being used to make control joints. New, improved saws have arrived on the market in recent years that make this technique fast and efficient. A sawed joint should be the same depth—¼ the slab thickness and a minimum of 1 inch deep—as a hand-tooled joint.

Saw the concrete as soon as it is firm enough to be worked without tearing the surface. Proper firmness typically occurs from 4 to 24 hours after the concrete is placed, but weather conditions may affect the waiting time. A slight raveling of the aggregate is acceptable and shows that the sawing is being done at about the right time.

Use an abrasive, non-diamond saw blade on green concrete that is set but not appreciably hardened. The loose particles in green concrete may tear the diamonds off a diamond blade.

Diamond blades are preferred for harder concrete and concrete with very hard aggregate. Use water for cooling when cutting with a diamond blade to avoid ruining the blade.

Using a Groover

Some builders use a grooving tool to create a channel in the surface of the concrete. (While not strictly a control joint, a groove performs something of the same function if it is deep enough. To create a groove using hand tools, use a trowel to dislodge aggregate along the path of the groove as soon as possible after bullfloating. Then use a special grooving tool to score the surface of the concrete. Some builders use the groover just before floating and troweling and again afterwards. Do not work on the concrete when bleed water is present.

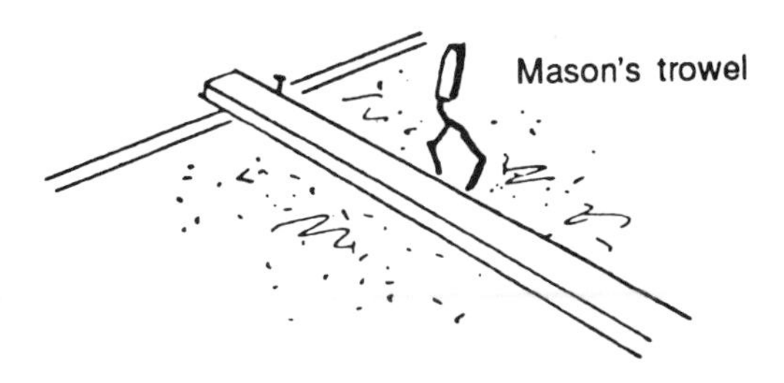

Using a trowel soon after bullfloating to dislodge aggregate makes using the groover much easier later.

Floating and Troweling

Troweling compacts, densifies, and hardens the surface of a concrete slab, making it smoother and more wear resistant. On industrial floors with a lot of traffic, the slab might be troweled several times. For a typical residential floor slab, a floating and troweling followed by a second troweling may be enough.

Concrete need not be troweled after floating if a smooth finish is not required. Sometimes the troweling should be omitted on areas such as driveways and sidewalks that do not need a dense surface. In such cases, use a dry broom to get the desired texture.

When to Start

Wait to hand float and trowel until the concrete is firm enough so that a person's weight on it makes no more than a ¼-inch imprint. The concrete should be a bit firmer if you plan to use a power float and trowel.

Bleed water should not be present on the surface. But lack of bleed water is not a sufficient test of readiness since air-entrained concrete and low-slump concrete may not bleed much.

In some cases the concrete starts to set up with bleed water present. In this situation the concrete can be finished but special measures must be taken.

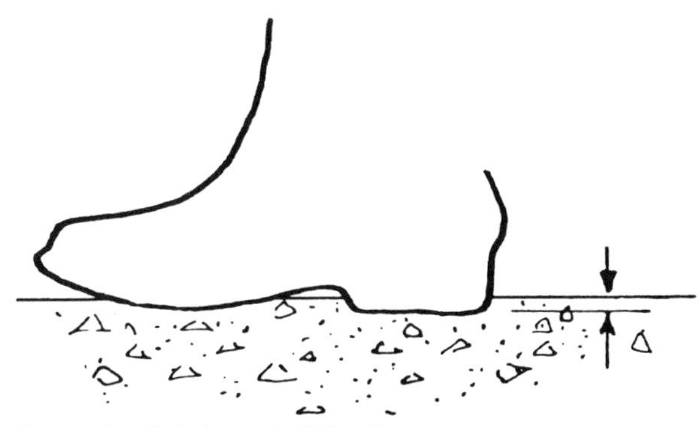

Footprint left by adult in the concrete should be no more than ¼ inch deep.

Sprinkling water on the surface to make finishing easier is a very risky practice because it may have the same effect as working the surface with bleed water present: it can weaken the surface and cause dusting, scaling, or crazing. Fogging or misting is a less risky technique because it is less likely to result in water accumulating on the surface of the concrete.

Hand Floating and Troweling

Start with floating and then trowel flat. Use a smaller trowel for subsequent trowelings, and increase the angle slightly each time. Keep the angle small—never increase it more than ½ inch. If the angle is increased too much, a washboard or chatter effect leaves ripples that are very hard to remove; and even after the ripples are removed, pinholes may remain. Pinholes are formed when sand grains are moved slightly on the surface of the concrete.

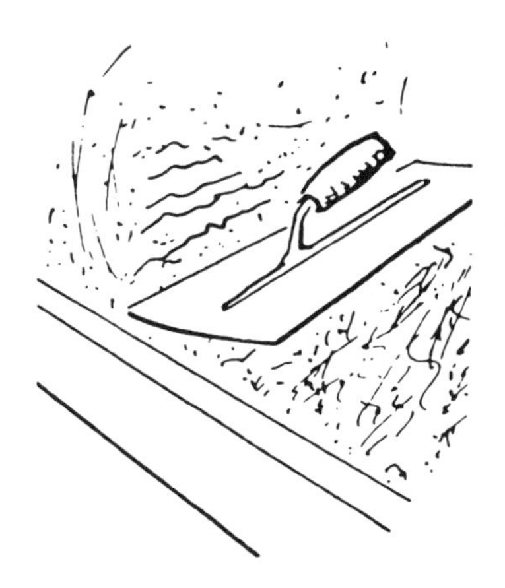

Too great a troweling angle leaves chatter marks that are hard to remove.

Use a wood float on concrete that is not air entrained or has high slump. Wood opens up the concrete, letting entrapped air escape, thereby helping prevent blisters. A wood float also makes a rough texture that allows better evaporation. Use a magnesium float on air-entrained concrete. Wood may tear the surface too much.

Do not over-trowel. Over-troweling can leave burn marks on the concrete, particularly if a steel trowel is used. Calcium chloride in the concrete may worsen this effect. Be particularly careful with white or colored concrete. Masons sometimes use plastic trowels or dampen the trowel slightly to help avoid problems.

Power Floating and Troweling

The concrete should be a bit harder before power floating than it is for hand floating. It is ready for power floating when an adult's weight makes no more than an ⅛-inch indentation in the surface.

Follow the same general blade-angle guidelines recommended for hand floating and troweling when a power trowel is used. Some experts recommend that the final troweling be done by hand, even if the slab is initially power troweled. Hand troweling can provide a superior final finish. Either way, hand troweling is required to finish corners and other tight spots that a troweling machine cannot reach.

Finishing Stairs

Early Stripping Method

Strike off and rough-float the top stair tread or landing. Edge, then hand float and hand trowel when the concrete has set up enough to support an adult's weight and leave an impression no more than ¼ inch deep.

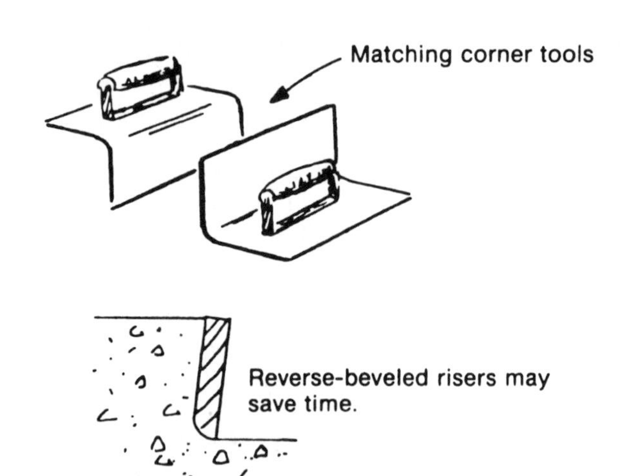

Continue down the steps. Float each tread and edge with a ¼- to ½-inch-radius tool, then trowel.

Wait until the steps have set up enough that they hold their shape when the riser boards are removed. The length of time varies depending on conditions, but initial set is usually 1 to 3 hours. Do not wait too long since it is easier to finish the risers before the final set of the concrete. Judging when to start finishing the risers can only be gained through experience.

When the steps have reached their initial set, remove the top riser form. Finish each riser before going down to the next step.

After troweling, draw a brush across the riser and the tread to produce a uniform-looking nonskid surface. This procedure is similar to lightly brooming a slab.

Move down, remove the next riser form, and finish that step. Work quickly so that the lower steps can be finished before the concrete sets too hard.

The sidewalls can also be finished the same day. First float and then finish with a ⅛- to ¼-inch layer of stucco. The mix should be 1 part portland cement to 1 ½ parts fine sand. The stucco should be spread with a trowel and then floated with a cork or sponge-rubber float. The surface may be troweled, brushed, or swirled. As an alternative, sidewall forms may be left in place several days for better curing. A grout clean-down, or "dressing," may be needed to give the final finish.

Late Stripping Method

Leaving the forms on the stairs for a few days will protect the fresh concrete, improve curing, and provide some insulation in cold weather.

Float, edge, trowel, and brush tread surfaces while the forms, including risers, are still in place. Correctly finishing near the bottom of riser forms is very important to avoid later trouble.

After removing each form, chip or grind off with a handstone all small projections. Any honeycomb areas should be chipped out and patched with a stiff mortar. Honeycomb will be minimized if care is taken in forming, placing, and consolidating the concrete. If risers and sidewalls are not uniform in color when the forms are stripped, a grout dressing can be used.

Grout Dressing

Finished concrete surfaces may have blotches, a slight film, mortar stains from leaks in the forms, or small pits caused by air bubbles. Grout dressing can camouflage these imperfections.

After defects have been repaired, saturate the surface thoroughly with water. A grout of 1 part portland cement and 1 ½ to 2 parts fine sand should be applied uniformly by brushing, spraying, or rubbing on with a burlap or sponge-rubber float. The mixture should completely fill small voids in the surface. White portland cement is often used for about one-quarter of the cement in the grout to give a lighter color. Float the surface vigorously with a rubber or wood float immediately after applying the grout. Then rub the surface with clean, dry burlap. No visible film of grout should remain on the surface after the rubbing. Rewet and rerub the surface if necessary. Grout should not remain on the surface overnight as it will become too difficult to remove. This work should be done in the shade and preferably in cool, damp weather. Curing should then be continued for 2 days.

Placing and Finishing in Hot, Dry, or Windy Weather

Concrete sets much faster in hot weather. Evaporation is also much faster in hot, dry, or windy weather. Water evaporates nine times faster in a 25-mph wind than it does with no wind. Losing control of concrete is easy under such conditions, and therefore special techniques are required, including changes in mix design and admixtures. The mix may need to be cooled, and special curing procedures used.

Plan ahead. When the concrete arrives, no time can be wasted. The forms and subgrade must be prepared, tools and workers standing by, curing materials at hand, and windbreaks and sunshades constructed if they are needed.

Plan the best time of day to place the concrete. In hot weather the best time may be very early in the morning or late in the evening.

Dampen the subgrade well in advance of placing the concrete. In some hot and dry areas of the country, the subgrade is dampened for 2 or 3 days in advance of placing. More typically, dampening is done the night before. A damp subgrade keeps water from being drawn too rapidly from the underside of the concrete. However, water should not be standing on the subgrade when the concrete mix is placed.

Be sure the mix is modified appropriately and is not too hot. Concrete temperature can be reduced by adding ice to the mix water at the ready mix plant. However, doing this adds to the cost. Another simple method to lower the temperature of the fresh concrete is sprinkling aggregate stockpiles. Evaporation lowers the temperature of the concrete; however, the ready mix concrete supplier must account for the additional moisture in the mix.

After placing and during finishing, evaporation must be controlled. Use windbreaks, sunshades, polyethylene sheets, wet burlap, or waterproof paper to cover the concrete between steps in finishing.

Begin curing immediately after finishing. During curing, take care to control the rate of evaporation to help prevent small shrinkage cracks from forming. (See "Plastic-Shrinkage Cracks" in Chapter 6.)

Placing and Finishing in Cold Weather

Just as heat speeds up hydration, cold weather slows it down. Concrete almost stops gaining strength when temperatures are near freezing. In cold weather, the concrete may freeze before it gains sufficient strength and watertightness. Good curing may help reclaim some strength, but special precautions should be taken when the air temperature is 40 degrees F or less, especially if the temperature is falling. Plastic shrinkage may develop in the concrete during cold weather, especially if the air is dry or if heat is used.

Guidelines for placing and finishing in cold weather include—

- Prepare in advance by ordering an appropriate mix and having items such as insulating blankets, heated enclosures, and admixtures at hand for an emergency.
- Do not place concrete on a frozen subgrade, or on a subgrade that is likely to freeze and draw heat away from the fresh concrete. (Also, a frozen subgrade is unstable: it may either heave or settle unevenly and crack the concrete.
- Try heated concrete, which is available from many ready mix plants located in cold climates.
- Insulate fresh concrete to trap heat and prevent freezing.
- Use high-early-strength cement and admixtures such as non-chloride accelerators to accelerate strength gain to help prevent damage due to freezing.

Special curing measures such as insulation and extra heating are needed in cold weather, especially during the first days after placement.

High-Early-Strength Concrete
Early strength gain is important for casting concrete in cold weather. Quickly finishing, starting curing, and providing freezing protection must proceed rapidly. Early-strength-gain concrete uses high-early-strength cement, or Type III, which is very similar to ordinary portland cement, or Type I, except that it is ground finer. This fineness causes the cement to hydrate more rapidly, thereby releasing heat and gaining strength more quickly. This greater heat helps protect the concrete from freezing temperatures while the concrete is gaining strength.

To avoid cold-weather problems, some builders stockpile foundations before cold weather starts, or they close in and heat basements before placing basement slabs.

Summary of Ordering, Placing, and Finishing the Concrete

Up-front attention to details when ordering concrete can prevent many problems later. Careful attention to scheduling, logistics, and the use of skilled workers helps ensure proper placement, curing, and a good finish. Use the following checklist as a reminder to help in purchasing and producing high-quality, durable concrete—

- Work with your supplier to be sure you receive the right concrete mix for the job.
- Order concrete to arrive at the jobsite at a specific time—and have the site prepared and workers ready for its arrival.
- Check routes of access for the ready mix truck to deposit the concrete into the forms near the final location.
- If ready access for the truck is not available, prepare to use alternate methods, like chutes, buggies, buckets, wheelbarrows, or a concrete pump as needed to transport the concrete quickly and efficiently once it arrives on-site.
- Have an adequate number of skilled crew ready.
- Check that proper placing and finishing tools are ready.
- Be sure that the crew has proper eye and skin protection.
- Anticipate what you will need to finish and cure the concrete properly, including special considerations for hot or cold weather.
- Strike off or screed the concrete level immediately after placing.
- Start finishing when the bleed water is evaporated or removed and the concrete is stiff enough to support the finisher's weight.

Chapter 5

Jointing, Reinforcing, and Curing

After building the forms, determining the placement of joints and reinforcing bars follows. Then, after concrete is placed in the forms and finished, careful curing of the concrete begins.

Concrete normally shrinks as it sets. The amount of shrinkage depends on the composition of the concrete and on the quality of curing. Careful layout and timely placement of control joints will cause the shrinkage cracks to form out of sight within the control joints instead of being randomly spaced.

Jointing

The three basic joints are control joints (also called contraction joints), isolation joints (also called expansion joints), and construction joints.

Control Joints determine where cracks form.

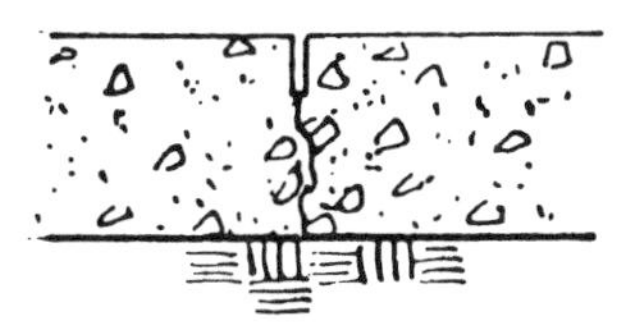

Isolation Joints allow relative movement of sections of concrete without causing cracks.

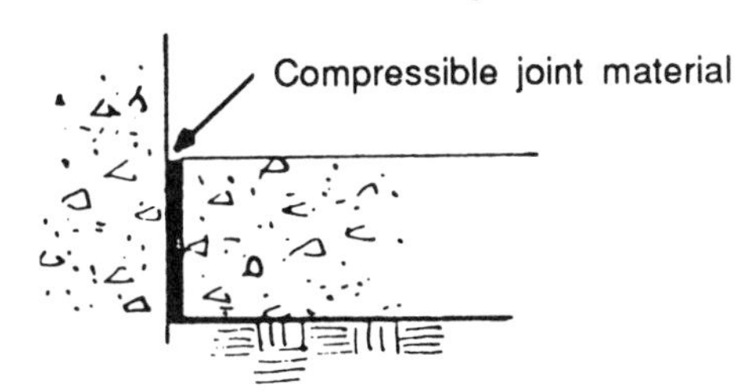

Construction joints connect hardened concrete from one casting to a later casting.

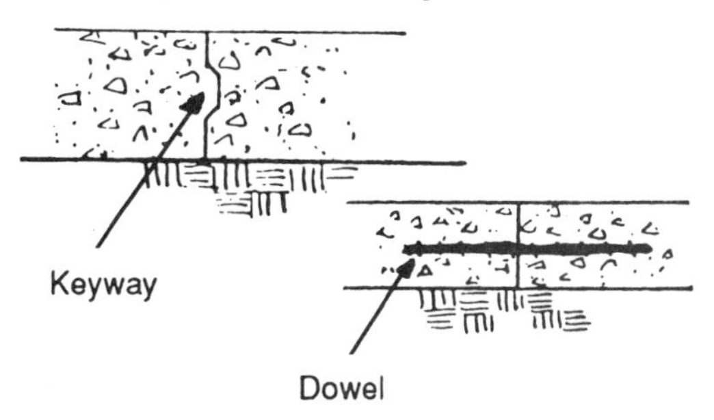

Control Joints

Control joints are placed where stresses accumulate and are likely to cause cracks. The control joint needs to be made deep—one-fourth the slab depth (and at minimum 1 inch deep)—to guide the direction of cracks when they form. If the control joint is too shallow, cracks may follow random patterns rather than forming in the joint as intended. On the other hand, if the control joint is too deep and aggregate interlock is lost, the concrete sections may not remain at the same level.

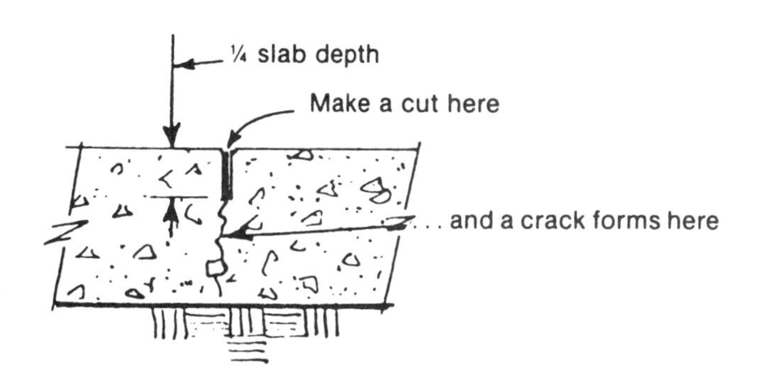

Locating Control Joints

Typically, control joints in slabs on ground should be spaced 10 to 15 feet apart, making concrete sections as nearly square as possible.

If the sections cannot be square, the longer sides should never be more than 1 ½ times the length of the shorter sides.

Assuming good curing, the required joint spacing can be influenced by many things, including—

- Total water in the mix. More water may mean more shrinkage, thus the need for more control joints.
- Weather conditions. High temperature and low humidity may mean more shrinkage and the need for more control joints.
- Aggregate. The type and size of aggregate may influence shrinkage and thus the spacing of control joints, as shown in the accompanying table.

Maximum Spacing of Control Joints, in Feet
Slump, 4"-5"

Slab thickness	Maximum-size aggregate: Less than ¾"	Maximum-size aggregate: ¾" and larger	Slump less than 4"
4"	8	10	12
5"	10	13	15
6"	12	15	18

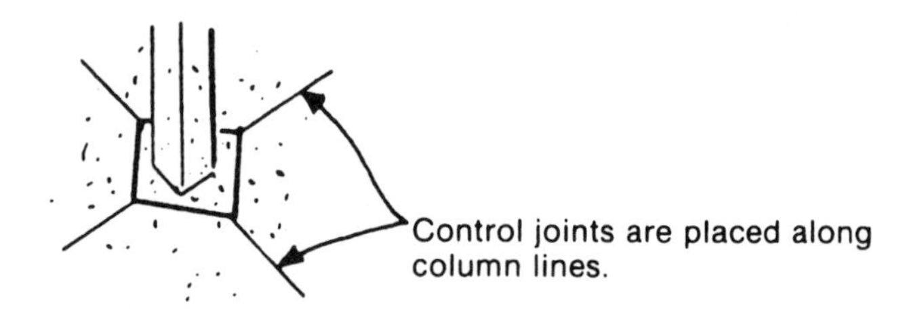

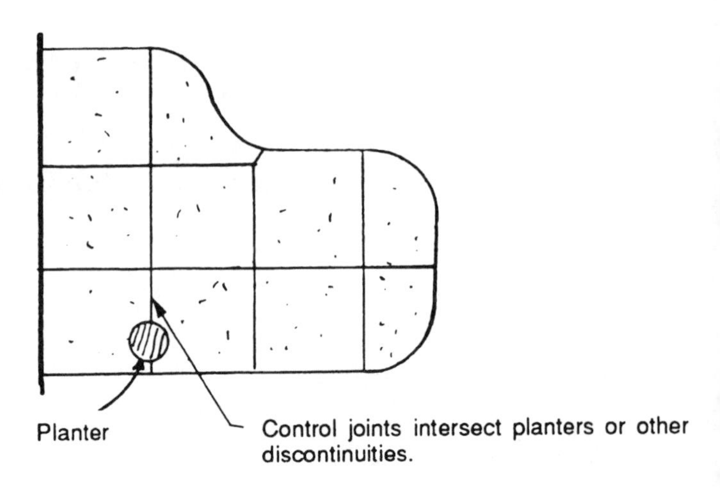

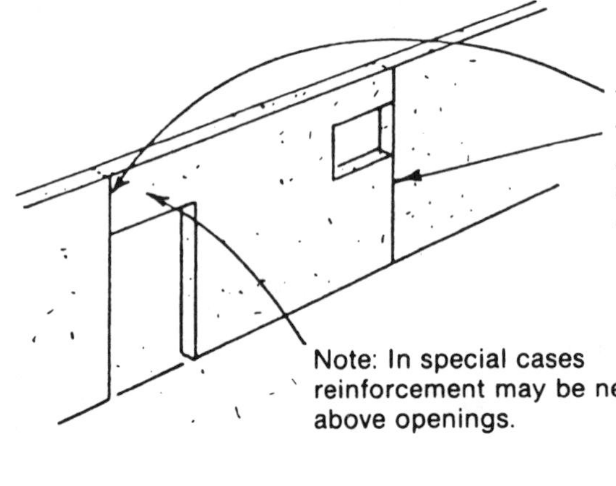

They may be placed in weak areas that are more likely to crack, such as in walls within 10 feet of corners and a maximum of 20 feet apart and at openings.

Note: In special cases reinforcement may be needed above openings.

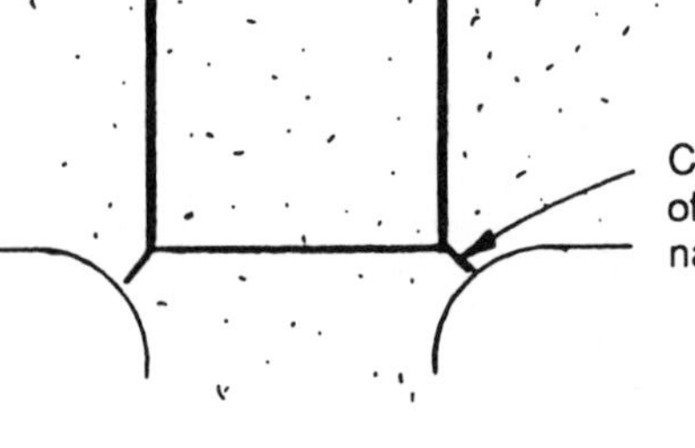

Correct method of jointing to avoid narrow sections

Control joints and construction joints are also required if a change of thickness occurs, possibly at a doorway or where heavy machinery is used.

Slope should not be more than 1 in 10.

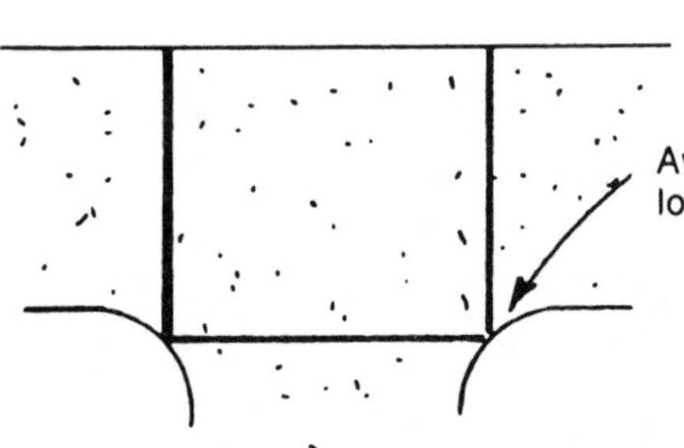

Avoid narrow sections when locating joints.

Control Joints for Topping Slabs

A bonded topping joint should be at least as wide as the base slab joint; it should be placed directly over the base joint, and should extend through the topping.

An unbonded topping should be free to move independently of the base; therefore, only the isolation joints need to line up. The topping joint should be at least half the thickness of the topping.

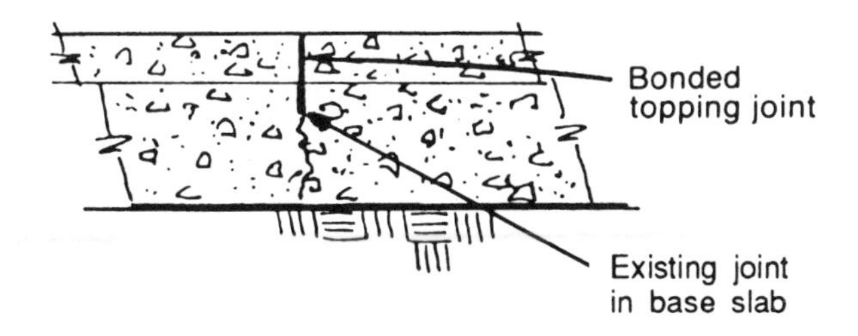

Control Joints for Curbs

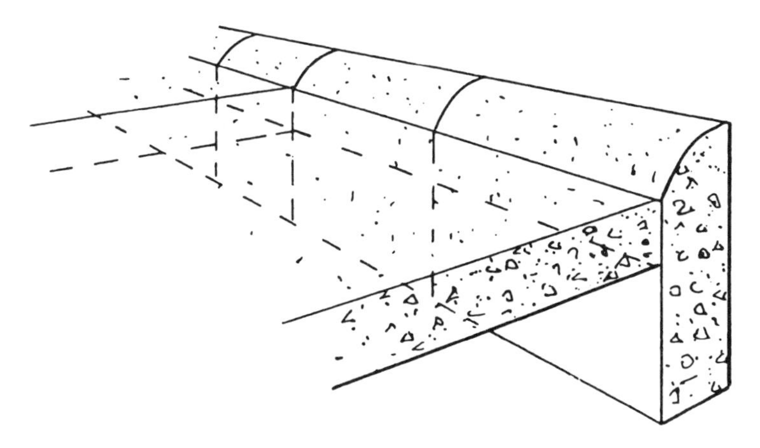

Continue the street-crack control joints through the curb, even though the curb may have additional joints.

Cutting Control Joints

For detailed guidelines on cutting control joints, including sawing joints, see "Making Control Joints" in Chapter 4.

First dislodge the aggregate.

Then use a special joint tool or a groover.

Divider strips of decay-resistant wood such as heart cedar, heart redwood, or treated wood may be used for control joints.

Sealing Control Joints

Seal control joints if they are apt to spall because of dirt, debris, or ice. Using an air-entrained mix, forming to the proper thickness, and proper finishing and curing also help to prevent spalling.

Many sealant materials are available, including polymers combined with coal tar, urethane, rubberized materials, asphalt combined with elastomeric polymers, wood, and plastics. The architect or the owner may specify the sealant. If it is not specified, the builder should consult local suppliers about the appropriate sealant to use.

Pointers on sealing joints—

- Proper curing is essential to proper functioning of control joints.
- Seal the joint flush with or slightly below the surface.
- Delay sealing as long as possible to allow the concrete to gain strength and to allow the joint to open up as much as possible.
- If you use a liquid curing compound, be sure to select one that is compatible with the joint sealant.
- Use compressed air or a vacuum to clean the joint before sealing.

Structural Slabs and Control Joints

Structural slabs contain 2 to 4 percent reinforcing steel and are typically unsupported underneath. These slabs must be continuous between supports and should not have control joints. They may develop fine cracks, but the steel is designed to keep these cracks from spreading.

Occasionally, a structural slab is built on grade, perhaps in a flood-prone area or on unstable soil. An engineer may or may not specify joints in such a situation, depending on the design.

Wire mesh in the usual amounts does not create a structural slab, and does not eliminate the need for control joints. If used, the mesh should not lap at control or isolation joints, but may extend through construction joints.

Basement Slabs and Control Joints

Control joints are often omitted in basement slabs except when required by code. The basement area is usually small and at a relatively uniform temperature, so major cracking is less likely. Still, cracks do occur, especially if a slab is irregularly shaped, and control joints will make the cracks less noticeable. If the slab is to be covered with tile or carpet, control joints are not needed.

Isolation Joints

Use isolation joints between concrete sections that need to move relative to each other.

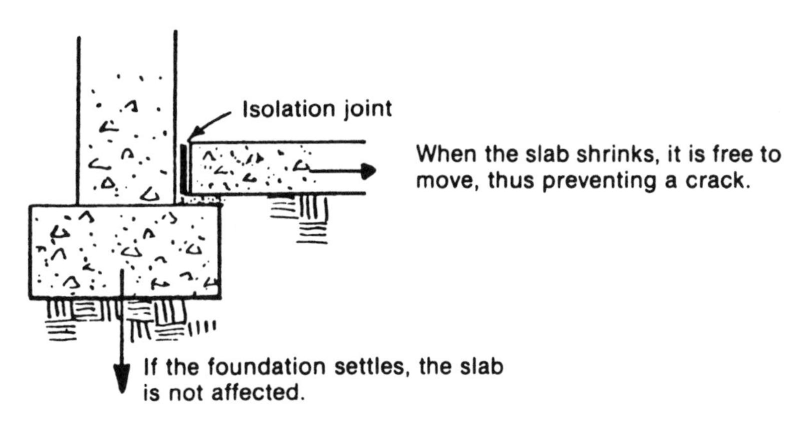

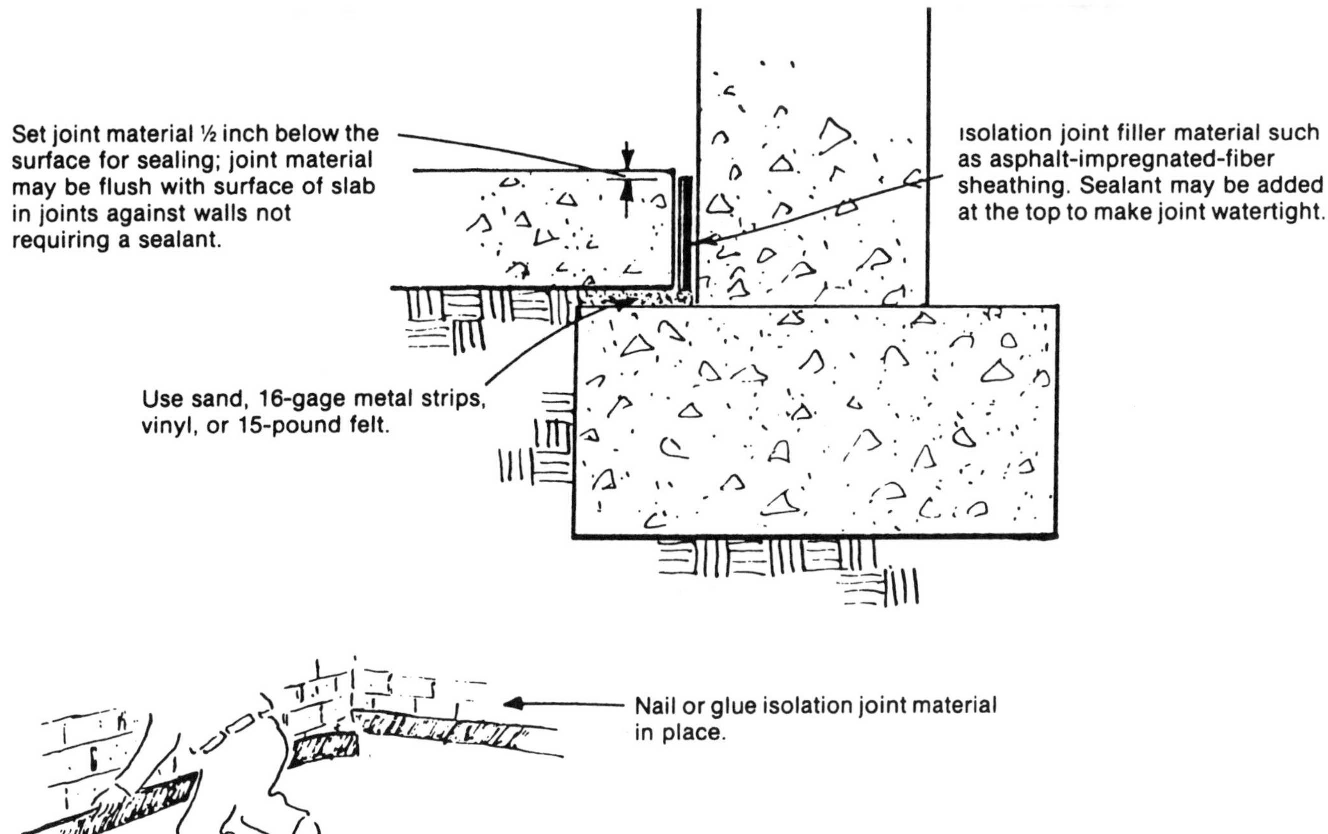

An Alternate Way to Form Isolation Joints

Wedge during construction. . .

. . .add joint sealer afterward.

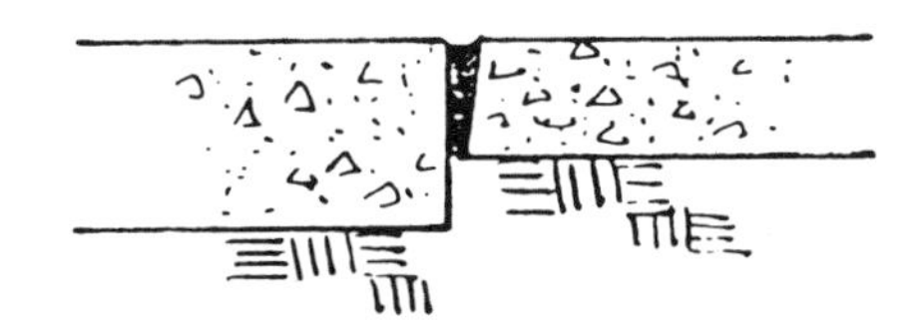

Other Examples of Isolation Joint Locations

Isolation joints are also needed around rigid objects such as pipe columns, drains, fireplugs, manholes, and utility poles.

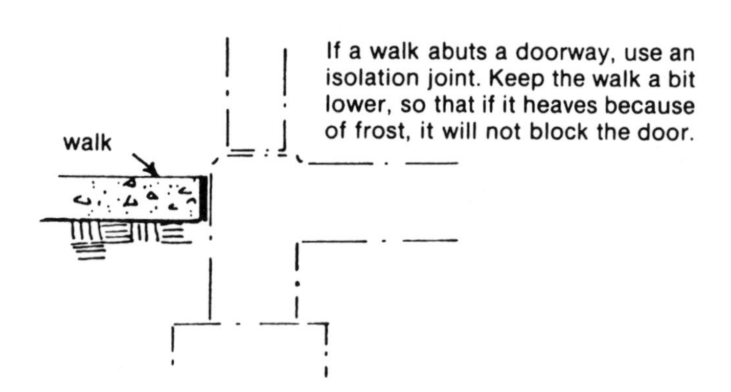

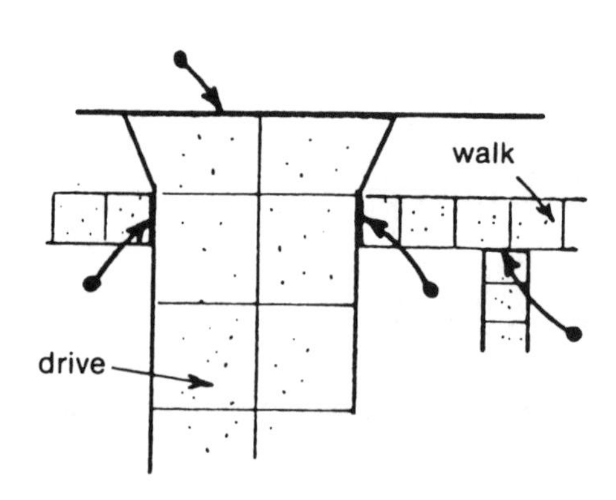

Arrows indicate isolation joints.

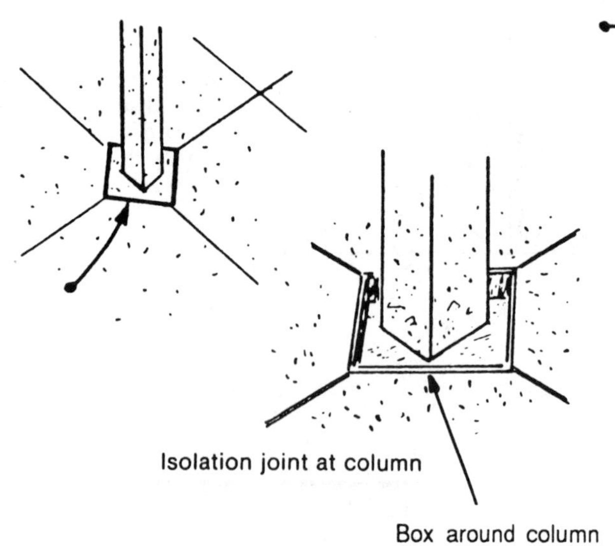

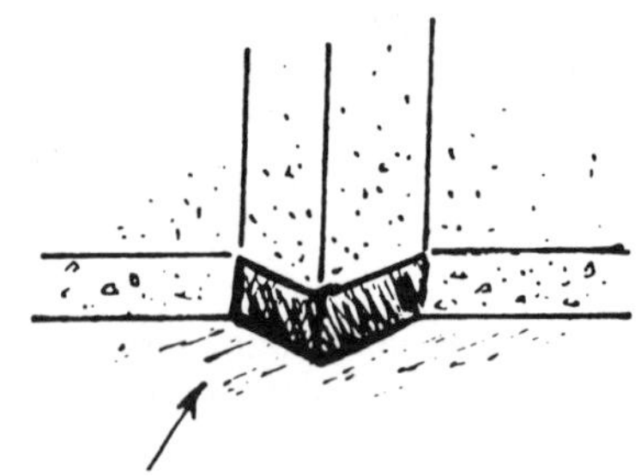

No box-out is needed if floor is to be covered with tile or carpet.

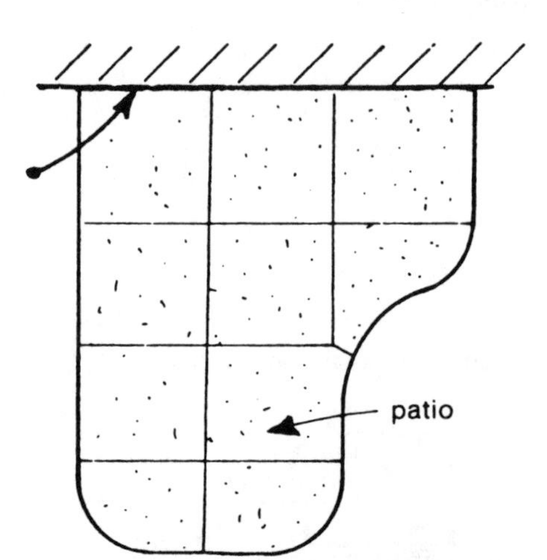

Arrows indicate isolation joints.

Construction Joints

Sometimes concreting must be interrupted—for example, at the end of the day—and a joint is required to tie the slab to the next day's placement. This is a construction joint.

Though a true construction joint is bonded so that it does not move at all, bonding usually is not necessary. Instead, the construction joint is made to work as a control joint, allowing horizontal but not vertical movement. A bond-breaker must be used at the joint to be sure it can move. An unbonded construction joint should be placed at a control joint location.

To make a true, rigid construction joint that does not move in any direction, steel reinforcing bars—called tie-bars—may be used.

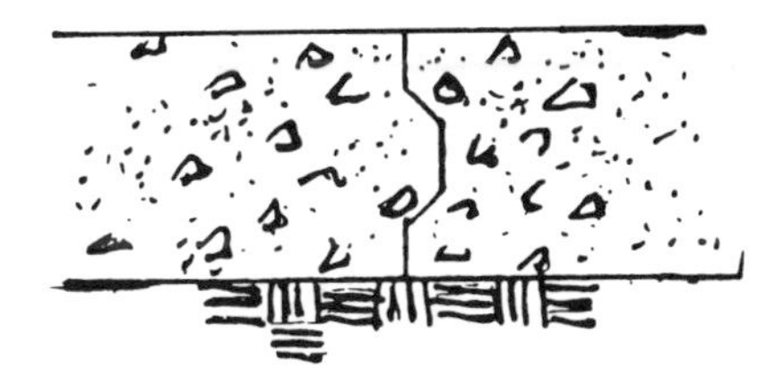

A typical construction joint is keyed so that neither side rises up past the other, and yet each side is free to move horizontally. This construction joint also acts as a control joint.

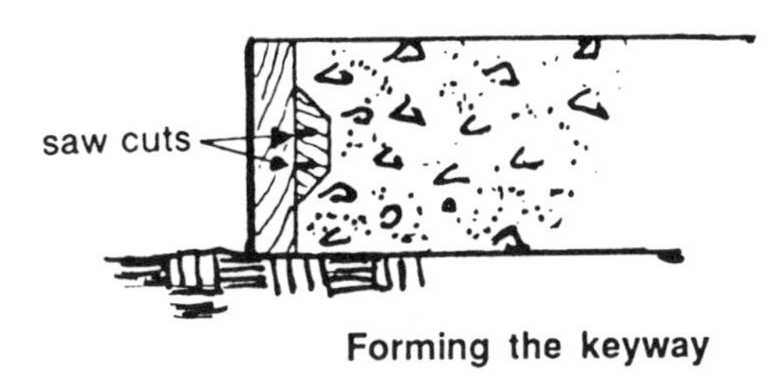

Forming the keyway

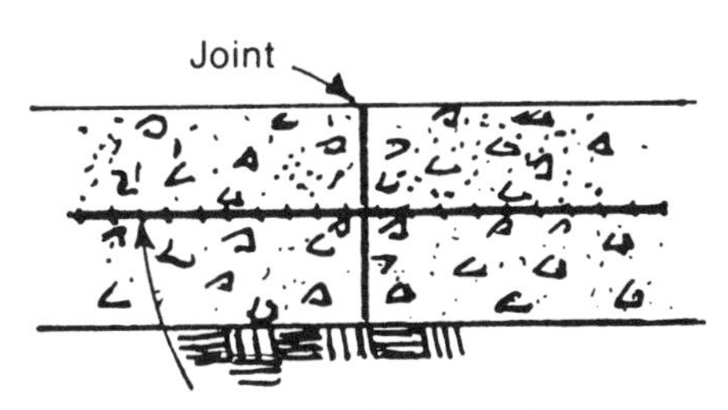

Reinforcing tie-bar to keep the crack from separating

Installing Reinforcement

Wire Mesh

A good method for placing mesh is to place the concrete slab to about half the desired thickness, strike it off, place the mesh, and then place the remaining concrete and strike it off. This method allows good control of the location of the mesh and keeps the mesh clean, but it is more costly in time and labor.

Another good method is to use concrete, steel, or plastic supports to hold up the mesh at the correct height. Do not use broken brick as a support; brick has a high level of water absorption, which can cause cracks to form in the concrete. Bricks also provide an entry point for moisture beneath the slab to corrode the wire mesh.

The need for mesh in slabs on grade is often questioned, although it may have value if it is sized and placed properly. Because mesh helps prevent cracks from spreading wider and keeps the aggregate interlocked at cracks, it helps prevent displacement—which can be important in meeting the standards of some warranty programs. Some codes require the use of mesh.

Where to Place Mesh

According to the Wire Reinforcement Institute (WRI), wire mesh should be placed 2 inches down from the surface to exert good control over cracking. Thus, it would be placed at the middle of a 4-inch slab. (Unlike wire mesh, fiber reinforcement is mixed into the fresh concrete and so is distributed throughout the concrete slab.) The use of wire mesh does not eliminate the need for jointing.

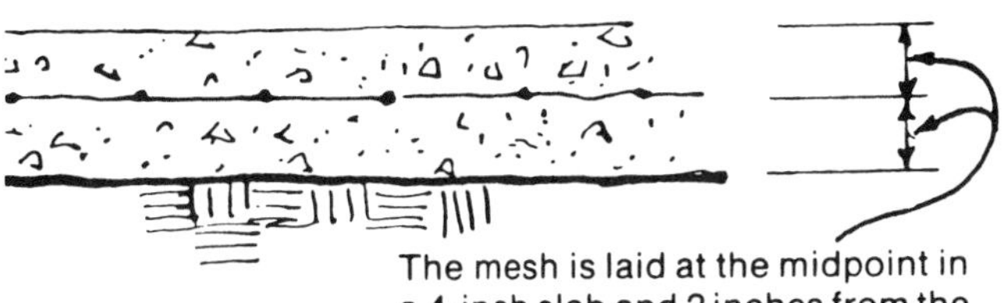

The mesh is laid at the midpoint in a 4-inch slab and 2 inches from the top in 5- and 6-inch slabs.

Control Joints and Mesh

Slabs with wire mesh require control joints, but the joints should not coincide with the lap of the mesh. If typical 6x6—W1.4xW1.4 mesh is used, stop the mesh several inches from each side of a joint. The joints should not be more than 15 feet apart.

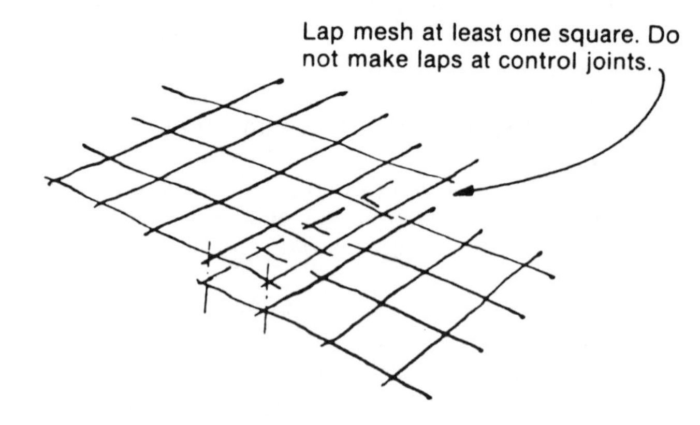

Ductwork and Mesh

Even if a floor does not contain wire mesh elsewhere, the area over ducts should be reinforced. A 6x6—W1.4xW1.4 fabric may be used and should be extended 18 inches past the point where the thickness of the slab returns to normal.

Size of Mesh

The Wire Reinforcing Institute (WRI) recommends 6x6—W1.4xW1.4 mesh for sidewalks and patios. It also recommends that wire spacing not exceed 6 inches in slabs that are less than 6 inches thick.

Reinforcing Steel

Cover for Reinforcing Steel

When reinforcing steel is used in concrete placed directly on earth or fill—for example, footings or a slab placed on earth—the minimum cover should be 3 inches, as shown in the sketch.

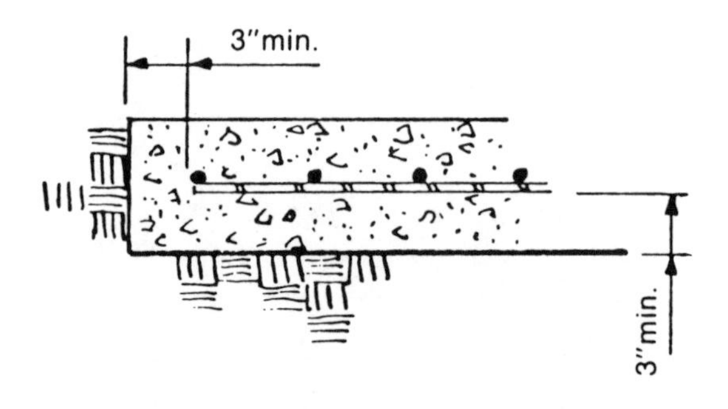

For concrete that is to be exposed to earth or weather, but not placed directly on earth, use 1 ½ or 2 inches of cover, depending upon bar sizes as follows—

Reinforcing steel	Cover
#6 through #18 bars	2 inches
#5 bars, ⅝-inch wire and smaller	1½ inches

For concrete that is not to be exposed to weather and will not be in contact with the ground (such as slabs, walls, and joists), use—

Reinforcing steel	Cover
#14 and #18 bars	1½ inches
#11 and smaller	¾ inch

Lapping Reinforcing Steel

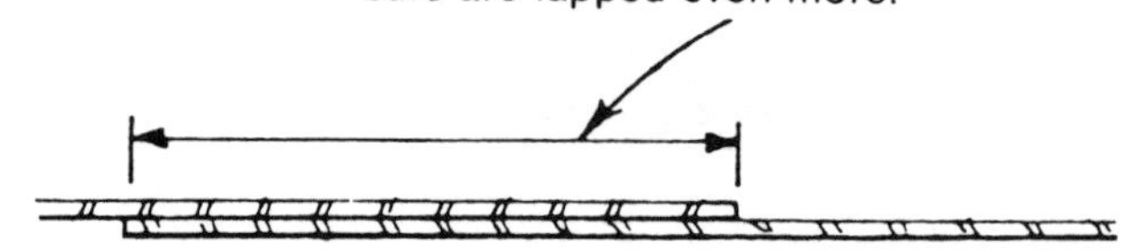

Curing

Curing is crucial to good concrete. Without a good cure, concrete dries out and may reach only 40 percent of its design strength.

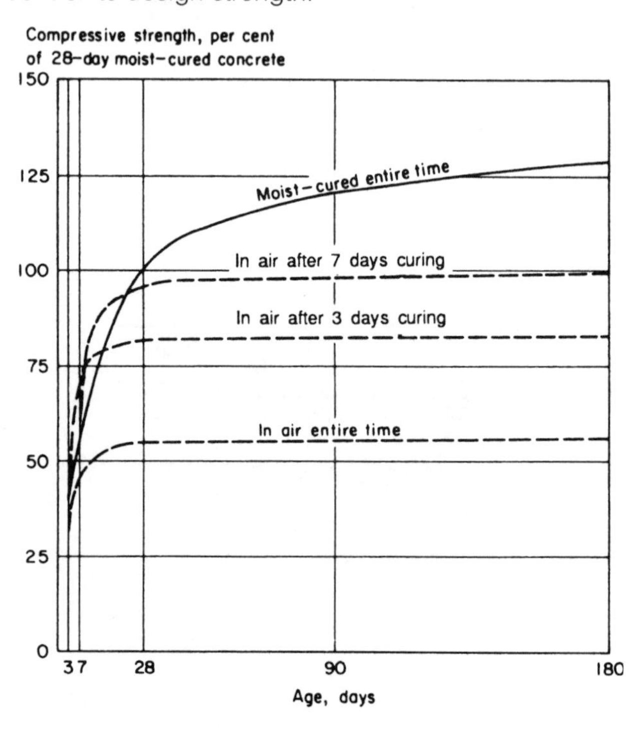

Concrete strength increases with age as long as moisture and favorable temperatures are present. Good curing more than doubles the strength of concrete.

As the chart shows, concrete continues to gain strength as long as it is curing. However, strength gain essentially stops soon after the curing is ended after 3 or 7 days.

All methods of curing keep the water in the concrete from escaping by either (1) sealing the surface without using water or (2) using water to seal the surface.

Curing occurs only with—

- water,
- a favorable temperature, and
- time.

Poorly cured concrete is more likely to crack and curl. Poorly cured concrete does not wear well, is less watertight, does not stand up to freezing and thawing or deicers, tends to have more efflorescence, and is more likely to be discolored.

Enough water for curing is already in the concrete no matter how low the water-cement ratio. Retaining this moisture is the most important part of curing, whatever the curing method used.

Normal air temperature supplies enough heat for curing. A good rule-of-thumb to follow is that if people are comfortable, the concrete is "comfortable" and continues to cure. However, when the temperature is near freezing, curing almost stops. Therefore, curing procedures must continue much longer at low temperatures, unless special precautions are taken.

The basic cure takes a minimum of—

- 5 days at 70 degrees F or higher, or
- 7 days at 50 to 70 degrees F.

Thin sections of concrete should cure longer. A longer cure will also result in more watertight concrete.

The lengths of time recommended for a basic cure are bare minimums. Longer curing should be allowed whenever possible. The most important curing time is the first few days. Begin curing as soon as the finishers get off the slab. Other pointers for good curing include—

- Take special care with edges and joints where concrete is likely to crack first. Inadequate curing is also likely at these points.
- Remember that if the joints are to receive sealants, or a slab is to be covered with linoleum, tile, or carpet, the cure must leave a surface that can be bonded.
- Keep the cure as uniform as possible in moisture, heat, and length of time. The cure should also be uniform in coverage—that is, do not cure some parts longer than others. For example, if a polyethylene sheet used for curing is ripped or blown away from a slab edge, uneven curing will result.
- Do not let a wet-cured concrete surface dry too rapidly. Taper off the curing. The same rule applies if heat is used—lower the temperature gradually (a few degrees per hour).
- After curing let the concrete surface air-dry for a few days before using it, unless a curing compound is applied. With a curing compound the concrete can be used soon after application, provided loads are low and the surface is protected from construction traffic. Always check manufacturers' recommendations.
- Longer air-drying is required if the concrete is air-entrained and is to be exposed to deicers; in this case, about 4 weeks are recommended. Curing compounds that do not allow air-drying should not be used when cold weather is approaching.

Curing without Adding Water

Polyethylene Cure

For a polyethylene cure, wide sheets of polyethylene are used to seal the surface. The sheets may be clear, white, or black. White is good in hot weather because it reflects light and heat. Black is good in cool weather because it absorbs heat.

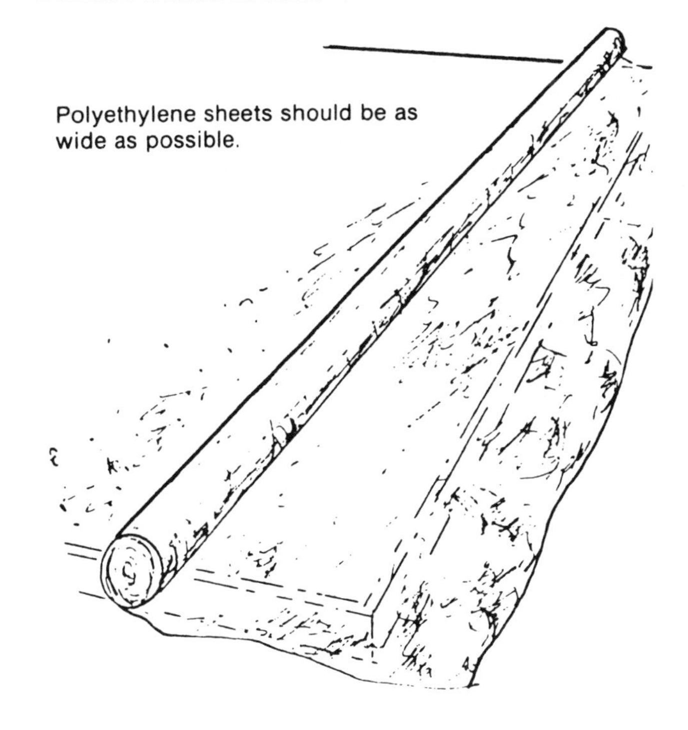

Polyethylene sheets should be as wide as possible.

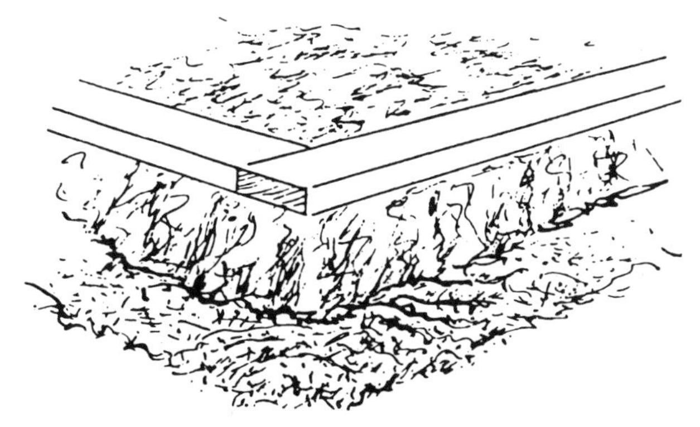

Carefully seal edges of polyethylene or waterproof paper and all joints with sand, tape, or mastic.

Wrinkles in polyethylene sheets produce mottled, discolored areas because of their greenhouse effect.

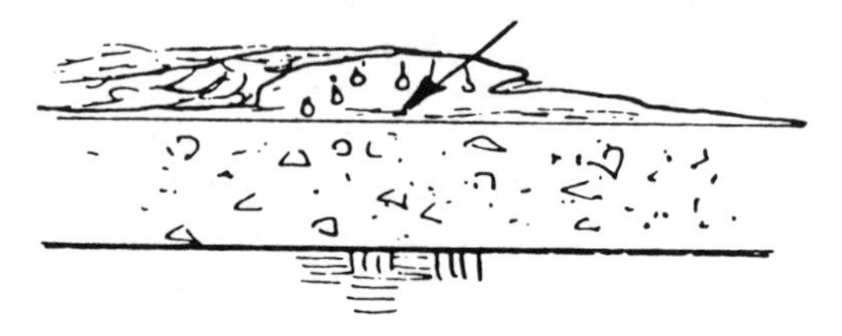

Wetting the concrete before applying polyethylene is a good idea. The edges and laps of the sheets must be sealed because vapor pressure builds up underneath and must be contained for a good cure. The edges can be sealed with lumber and sand or earth. Joints can be sealed with sand, nonstaining mastic, or tape. Polyethylene sheets used for curing should be as wide as possible to avoid unnecessary laps.

Advantages and Disadvantages. Polyethylene is easy to handle and place except in windy weather. It does not leave a film that will prevent a bond with floor coverings, and it provides some protection against stains and debris. It is low cost, particularly if it is taken care of and can be reused, and it does not rot or mildew. It can be used for other purposes afterward, such as protecting lumber.

On the other hand, polyethylene that is placed in areas with a lot of activity can be punctured easily or the edges or laps can open. If this happens, the cure will be uneven and stresses and cracking may result.

Sometimes polyethylene leaves a blotchy or mottled surface. This condition, called the greenhouse effect, is caused by water vapor condensing under wrinkles in the film and running down into puddles where the film is in contact with the slab. The damp spots change the color of the concrete slightly.

Because of the greenhouse effect and the potential for discoloration, polyethylene curing is not recommended for colored concretes. Theoretically, if the film is perfectly flat, the greenhouse effect does not occur. In practice, the film is almost certain to wrinkle.

Waterproof Paper

The waterproof paper cure is quite similar to the polyethylene cure. Ordinary building paper should not be used; it may be water- repellent but not vaporproof.

Curing paper must be vaporproof, strong, nonstaining, and nonshrinking. Good curing paper withstands some abuse and is reusable, although protecting it from construction traffic is a good idea.

Use wide sheets, seal all edges with sand or lumber, and seal laps with nonstaining mastic, glue, or tape. Keep the paper flat and wrinkle-free.

Wet the concrete thoroughly before placing the paper. Light-colored sheets are available for hot-weather curing and dark sheets for cool-weather curing.

Advantages and Disadvantages. Curing paper leaves the concrete surface clean and able to bond well with a floor covering. It gives some protection from stains and debris. Good curing paper usually has fewer wrinkles than the thinner, stretchable polyethylene; thus blotching from the greenhouse effect is less likely.

Liquid-Membrane Curing Compounds

Among the many types of membrane cures, some only cure while others claim also to seal, harden, or dustproof. These products may have a wax, resin, asphaltic, acrylic, epoxy, sodium silicate, rubber, or other base. Builders should check the nature of the compounds and read their labels carefully before using them. Membrane curing compounds should meet the ASTM C 156 and C 309 standards. Builders seeking more information on ASTM standards should consult the information sources listed in the back of this book.

The compound may seal the surface for a month or so and then gradually dissipate or wear off, depending on the type.

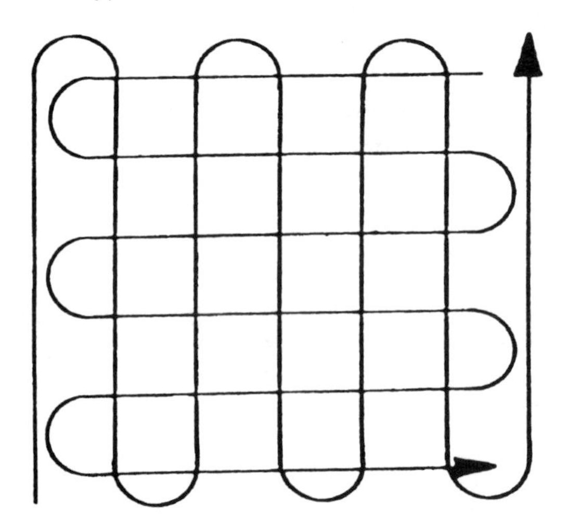

Spraying pattern

Good coverage is extremely important when curing compounds are used. Rather than make one heavy spraying pass, better coverage is obtained by making two lighter passes at right angles to each other.

A colored curing compound will allow missed spots to be seen right away. White, black, and grey liquid-membrane compounds are available. A fugitive (vanishing) dye can be added to clear compounds; later it will virtually disappear.

Another way to check for coverage when using a clear liquid membrane is to cover a portion of the slab with polyethylene or waterproof paper. Check it later to see if moisture is accumulating under the covering. If it is, the membrane coverage is not good, nor is the cure.

Clear Wax Base

A clear wax-base membrane curing compound is relatively inexpensive and is effective for about 28 days. This type of curing compound is not recommended if the concrete is to be painted, tiled, or treated since the wax base leaves a slightly gummy residue that collects dirt. If a decision is made to paint a wax-cured surface, it may be necessary to wait a year or more until the wax membrane fully disintegrates.

Clear Wax-Resin Base

The wax-resin cure is one of the most commonly used today. It is typically less expensive than a resin cure. However, the waxy residue makes a poor surface for painting, tiling, or other surface treatment until it disintegrates. The length of time required for disintegration may be from 2 to 9 months, depending on the product. Painting over a wax-resin base cure is not recommended until more than a year after the compound is applied, but again this depends on the product.

A fugitive (vanishing) dye can be used to show coverage.

Clear Resin Base

The resin-base cure is another popular cure. It may be more expensive than a wax-resin cure, but it has the advantage of allowing painting or other surface treatment, depending on the product. Some resin cures may not allow surface treatment until they disintegrate, in 30 to 60 days or longer.

The resin cure typically results in a harder and more abrasion-resistant residual film than the wax cure, which is less likely to wear off under traffic.

A fugitive dye can be used with resin-base cures.

White-Pigmented Compounds

White-pigmented compounds may have wax or resin bases, but they are special formulations. Their advantage is their ability to lower heat buildup and evaporation in addition to indicating coverage.

White-pigmented cures often need to be stirred before using.

Black and Grey Compounds

Black and grey compounds with a complete or partial asphaltic base can be also used for curing. Asphaltic-base compounds are usually relatively inexpensive. They are good waterproofers, especially when sandwiched between layers of a two-course slab. They can also be used on the slab surface and provide a good base for tile, linoleum, and other materials compatible with asphalt.

Black or grey compounds absorb heat from the sun, a virtue in cold weather.

Other Membrane Cures

Other membrane cures include chlorinated rubber, acrylics, epoxies, and sodium silicates. They may harden, dustproof, and seal against oil, depending on the product. Builders should check manufacturers' guidelines before using. Some experts do not recommend sodium silicate compounds. Not all compounds used for curing or for concrete form release are environmentally safe. The Environmental Protection Agency (EPA) regulates the use of solvent-type materials. Be sure to check the manufacturer's instructions for any chemicals used at the jobsite.

Field tests show that synthetic-rubber-base compounds are far superior to resin-base compounds. The synthetic-rubber compounds may cost considerably more than resin cures, but experts believe they provide a superior cure.

When to Apply a Curing Compound

The best time to apply most curing compounds is right after finishing. Do not apply a curing compound when the surface has standing water or is otherwise too wet. Applying a compound under wet conditions will produce a poor surface seal.

On the other hand, if the curing compound is applied when the concrete surface has dried out too much, the compound will be absorbed into the concrete and the surface seal will be poor. If the surface has dried out, fog the concrete to seal the pores before applying a membrane compound.

Some of the cures that claim to seal, harden, and dustproof are applied at a later time—for instance, 8 hours after the concrete is finished. Check the manufacturer's recommendations.

In any case, apply the membrane compound as soon as the application is allowed, before the concrete dries out too much.

Joints and Membrane Cures

If joints are to be sealed with a joint sealant, be sure that no adhesion problem will exist because of the curing compound used. Use a curing compound that is compatible with the sealant, or cover the joints before applying the compound so that it cannot get inside them. If the compound is omitted at the joints, they must be cured in some other way.

Deicers and Membrane Cures

If concrete is placed in autumn and is expected to be exposed to freeze-thaw or deicers within a short period of time, a membrane cure may not be recommended since recently applied membrane compounds do not allow the 4 weeks of air-drying needed before subjection to freeze-thaw or deicers.

Advantages and Disadvantages. Membrane compounds are simple to apply, inexpensive both in labor and materials, and typically allow construction activity to proceed soon afterward. A cure is reliable if it meets ASTM C 156 and C 309 requirements provided it is applied properly and is protected when necessary.

White compounds are available that lessen heat and evaporation in hot weather. Black or grey asphaltic curing compounds help warm the concrete in cool weather and are good waterproofers, especially for two-layer slabs. They also provide good adhesion for floor-finish materials bonded with asphaltic adhesives.

Coverage can be a problem if clear membrane cures are used. If the coverage is poor, curing will be poor. For this reason, pigmented cures are advised. Fugitive dyes that later virtually disappear are available for clear compounds. Two thin applications at right angles to each other are recommended for good coverage.

Painting, sealants for joints, and other surface coatings can be a problem with membrane compounds. Wax-type compounds make adhesion particularly difficult. Check the compound for compatibility with any surface coating.

Deicers can cause problems for membrane-cured concrete if the curing takes place late in the fall. The curing material may not have a chance to disintegrate and allow the 4 weeks of air-drying needed before exposure to deicers.

Always check manufacturers' recommendations before using membrane cures.

Curing by Adding Water

Generally, curing with water is effective if it is done properly. The basic problem with wet curing is keeping the concrete continuously and uniformly wet. In some cases, an unreliable cure is worse than no cure at all.

Wet cures should be tapered off gradually. The concrete should air-dry for a few days before use, and for 4 weeks if it is to be exposed to freeze-thaw or deicers.

Wet Burlap

The proper burlap for wet curing is a specially treated material available in rolls. It is fireproof and rot-resistant and does not discolor or otherwise harm the concrete. Some burlaps are coated with plastic or aluminum to reflect light and heat.

The specially treated burlap is spread over the concrete and kept continuously and uniformly wet during the basic curing period. It must not be allowed to dry out. If it does, it may actually draw water from the concrete—and that will be worse than no cure at all.

Advantages and Disadvantages. Wet burlap is a very effective curing method if it is handled properly. It does not mottle or discolor the concrete surface, and it leaves a clean surface that can be painted or otherwise coated. Good burlap is economical and reusable. The principal disadvantage of wet burlap is the need to ensure that the burlap stays continuously wet. An automatic sprinkling system may break down and manpower may not always wet the burlap on time. The cost of labor and equipment to do the wetting also adds to the overall cost.

Another disadvantage of using wet burlap is that activity on the slab may have to be restricted during the cure.

Sprinkling or Fogging

A mechanical arrangement of pipes or hoses with sprinklers is stretched across or around the concrete when a sprinkling or fogging cure is used. The system may turn on automatically or be turned on by hand at set intervals.

The water may run off or evaporate more rapidly during hot, dry, or windy weather, and therefore the intervals between sprinkling will vary. If the surface dries between sprinklings, crazing or cracking of the concrete can result.

Advantages and Disadvantages. Like the wet-burlap cure, sprinkling or fogging is a good cure if done properly. It does not mottle or discolor the surface if it does not dry out in spots. It leaves a clean surface suitable for painting or other coatings.

A sprinkling or fogging cure may be relatively costly in terms of equipment, labor, and supervision. Possibly a greater disadvantage is the difficulty of maintaining quality control. The equipment may break down, the pipes may leak, the nozzles may clog, or the wind may blow and make some areas difficult to dampen. Low humidity, heat, and wind accelerate the evaporation rate, and dry spots may occur suddenly. Also, activity on the slab is impossible during curing.

Ponding

Ponding is probably the best curing method for quality concrete. Some sort of barrier such as a wooden frame must be placed around the concrete to contain the curing water, which may be several inches deep.

When the curing period is over and the pond is drained, the surface should be sprayed or fogged occasionally for a day or two, tapering it off gradually.

Ponding is not often used in residential concreting. Possibly it is thought to be too troublesome or too costly to supervise. These criticisms often are not justified. In fact, ponding can be both effective and economical. The areas to be cured in residential concreting are small, and sometimes they need little in the way of dams. If the system is designed properly, little supervision is needed.

Advantages and Disadvantages. Ponding may be the best cure for strong, watertight concrete. It is economical and requires little supervision if done correctly. It leaves a clean surface suitable for painting or any type of covering.

Ponding has an additional advantage: The water serves as a buffer or insulator against the environment and keeps the concrete at a more constant temperature regardless of exterior heat, cold, humidity, or wind.

A disadvantage of ponding is that activity on the slab is halted during the cure.

Quality control can be a problem, though probably not as much as with the wet-burlap and sprinkling cures. The damming system could break and the water be lost. Leaks around the edges of the dam could affect a poorly prepared subgrade. However, these problems can be avoided with good construction practices.

As with all cures, a supervisor should check regularly to see that no problems develop.

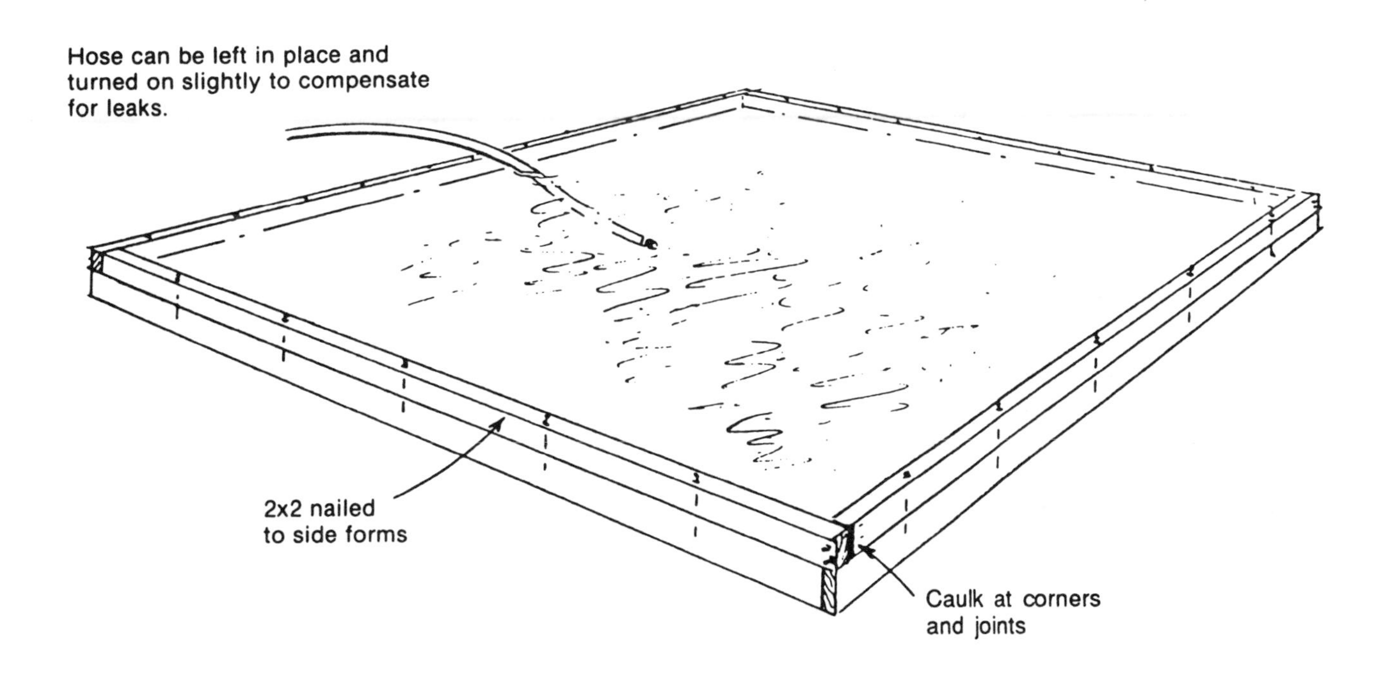

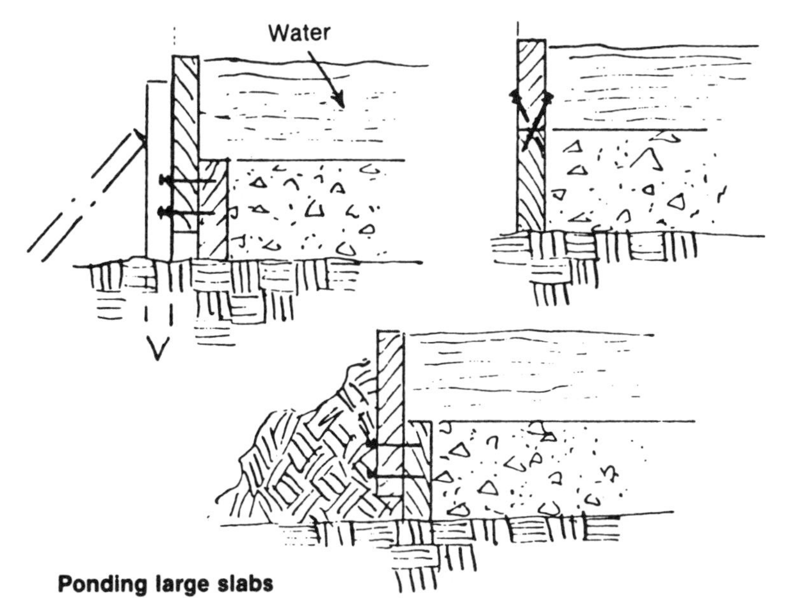

Ponding large slabs

Large slabs may need deep edge strips, since the slope of the concrete drops one side somewhat lower than the other.

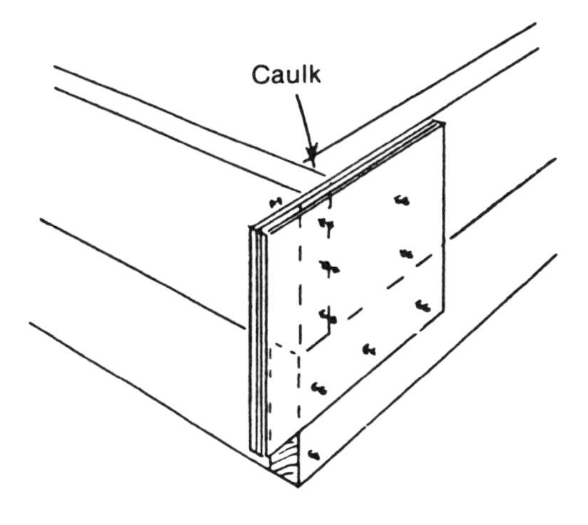

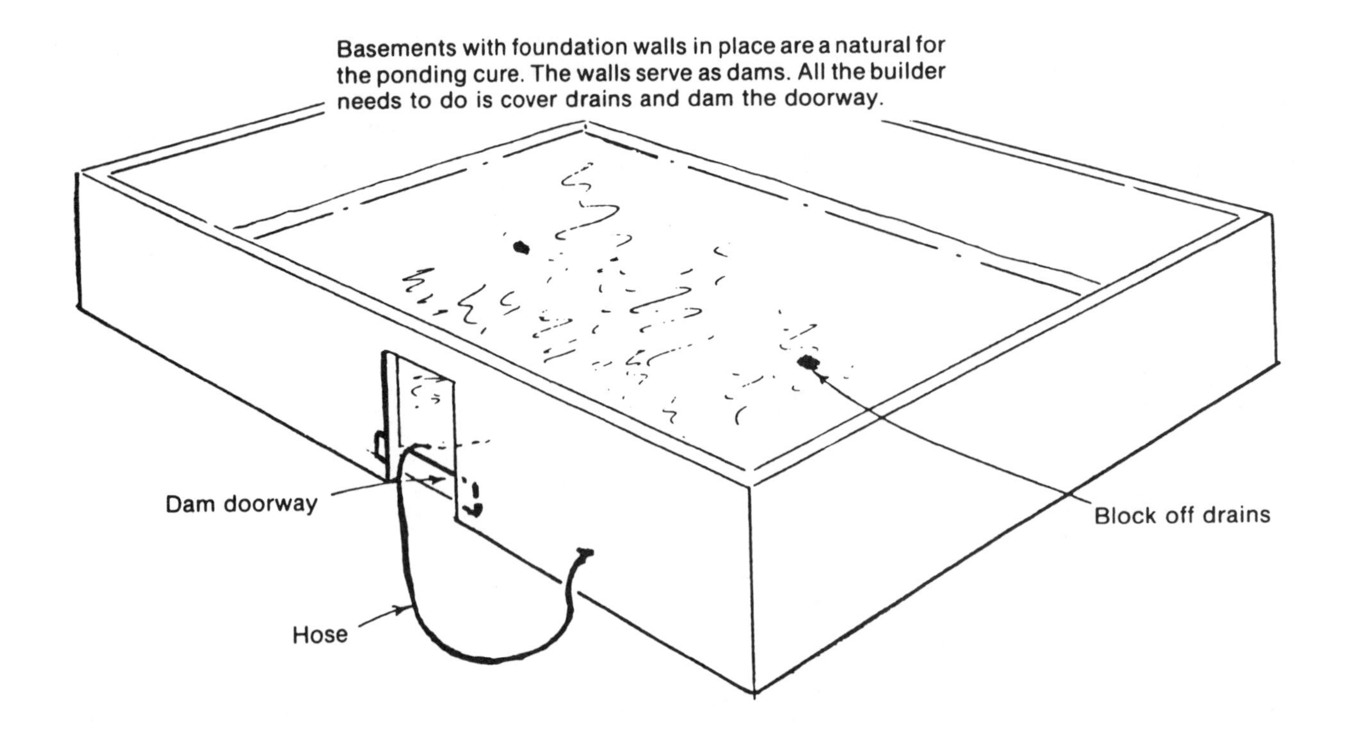

Curing in Hot, Dry, or Windy Weather

Special attention to curing is required when placing concrete in hot, dry, or windy weather. Beginning the cure immediately after finishing—although always important—is especially important in hot, dry, or windy weather. Measures to retain moisture in the concrete can be started during the finishing operation by using wet burlap, polyethylene, covers, shades, and windbreaks. Other measures that should be taken when placing concrete in hot, dry, or windy weather—relating to subgrade preparation, mix design, cooling, and evaporation control—are discussed in Chapters 1, 2, and 4.

Ponding may be the best curing method in hot, dry weather. It is not only an effective cure, but the water also serves as a buffer, keeping the concrete cool and at a more uniform temperature. Ponding may also be more reliable than other methods—less likely to break down or slip in quality.

No matter what type of cure is used in hot, dry, or windy weather, quality control is extremely important. Heat and rapid evaporation will encourage stresses and cracks unless moisture is retained with good curing techniques.

Cures that do not require additional water—such as polyethylene sheets, waterproof paper, or a liquid membrane—may be chosen if traffic is necessary on the slab during curing or because such a cure is more economical under the particular circumstances. When these cures are used, the slab should be dampened before curing is begun.

Whenever possible, use a white-tinted curing material to help control heat buildup and evaporation.

With polyethylene or paper cures, special care is needed to properly seal all laps and edges. This is particularly important because of evaporation in hot, dry, or windy weather.

Wrinkles in the polyethylene sheets may be more likely to cause mottling from the greenhouse effect in hot, dry weather.

In all cases, but particularly when wet curing is used in hot, dry, or windy weather, the cure should be tapered off by gradually reducing the amount and frequency of water applied.

Curing in Cold Weather

Cold weather requires special curing procedures and also special attention to subgrade preparation, mix design, protection, admixtures such as calcium chloride, and special finishing precautions. (See the sections titled "Accelerators" in Chapter 1, "The Subgrade in Cold Weather" in Chapter 2, and "Placing and Finishing in Cold Weather" in Chapter 4.) If concrete freezes before curing is completed, it can be permanently damaged. Resuming proper curing after concrete has frozen will help restore its desirable properties, but it may not restore all.

Faster curing can often save the concrete as well as save time. Use of high-early-strength cement is one way of accelerating hydration and thus curing. For plain concrete, the addition of calcium chloride in the mix is another way to speed the curing process; however, this method is not recommended for general use. Heating the sand, coarse aggregate, and water will also speed up hydration.

More important are measures to keep the concrete warm for at least several days after finishing. This greatly increases the rate of strength gain.

The concrete itself supplies some heat for curing; in cold weather the heat can be trapped in the concrete with an insulating material. Special blankets or two layers of polyethylene film or waterproof paper with straw between the layers are good insulators.

The edges of the insulating material must be firmly held down to keep the cure uniform so that cracking stresses will not develop from variations in temperature and moisture.

Very cold weather may require additional heat to be supplied to aid in curing the concrete. Do not use oil or gas heaters or any type of heater that gives off exhaust fumes unless the exhaust is vented to the outside. Carbon dioxide may cause the concrete to dust, and carbon monoxide may harm or kill people who breathe it.

The heat must be evenly dispersed to avoid cracking, crazing, or dusting. Heated air is low in humidity, so the concrete surface should also be wetted.

When insulating green concrete use bricks or lumber to hold down layered edges of polyethylene sheets that have straw sandwiched between the layers.

Vent heater exhausts to the outside to avoid harming concrete or workers.

After the cure, lower the heat gradually (several degrees per hour) so that the concrete does not experience thermal shock and crack. The temperature of the concrete should not drop more than 40 degrees F in 24 hours.

Protecting Concrete from Construction Traffic

As concrete is curing, it gains strength slowly; during this period, construction loads and traffic may exceed the strength and may mar the surface. Moreover, residential slabs, driveways, and walks even when they are mature are not designed to withstand heavy loads such as ready mix concrete trucks, lumber delivery trucks, or earth-moving equipment. These loads should always be restricted from residential concrete slabs.

If it is important to start constructing the building frame as soon as possible, smaller equipment, foot traffic, and some material storage loads may be acceptable on concrete that is 3 to 7 days old, provided the loads are low and the concrete surface is protected. Kraft paper, plywood sheets, or fiber sheets may provide adequate traffic paths over a concrete slab.

Protect Fresh Concrete from Loads

Plain concrete without rebar is strong in compression, but weak in tension. Therefore, loading that bends the concrete, especially at edges of slabs, basement walls, or retaining walls, will usually cause cracks. In case of doubt concerning loadings, common sense is the best protection for vulnerable new concrete.

Comparison of Curing Methods

Methods	Advantages	Disadvantages
Polyethylene	Easy to handle Inexpensive	Can be punctured May leave blotchy surface
Waterproof paper	Leaves surface clean	Deteriorates with traffic
Wax base, clear	Least expensive	Leaves gummy residue
Wax-resin base	Inexpensive	Leaves gummy residue
Resin base, clear	More expensive	Allows painting
White-pigmented	Reflects heat	Must be stirred before use
Black/grey mix	Asphalt tile sticks Inexpensive Absorbs heat	Leaves surface colored
Wet burlap	Economical, reusable Leaves surface clean	Must be kept wet Restricts traffic
Sprinkling, fogging	Leaves surface clean	Costly in labor Quality control difficult Restricts traffic Not good in cold weather
Ponding	Best quality control Economical Little supervision Leaves clean surface Temperature buffer	Pond may leak Restricts traffic Not good in cold weather

Summary of Jointing, Reinforcing, and Curing

Joints and reinforcements help control cracking from shrinkage as the concrete dries and also enhance the strength and durability of concrete elements. During curing, the ingredients in the concrete mix chemically interact, forming a durable bond. Some points to keep in mind about joints, reinforcements, and curing are—

- The depth of control joints should be carefully planned to control cracking while maintaining good interlock of the aggregate.
- Spacing of joints in slabs can be affected by aggregate size and water content of the mix, the shape of the slab, and even the weather.
- If joints are to be sealed, allow enough time before applying the sealant to let the joint cure—and expand—thoroughly.
- Use isolation joints between concrete sections that need to move relative to each other, and around rigid objects such as pipe columns or drains.
- Reinforcing mesh should be placed about 2 inches below the surface (for example, in the middle of a 4-inch slab).
- Reinforcing steel bars should have appropriate cover depending on their size, on whether the concrete is placed directly on earth or fill, and whether the concrete will be directly exposed to weather.
- Longer curing adds to the strength and watertightness of concrete and can be particularly important for thin sections of concrete.
- Keep the cure as uniform as possible.
- Plan ahead and select a curing method appropriate to the site conditions, including anticipated weather.
- Protect the concrete from construction traffic while it is curing.

Chapter 6

Concrete Problems

Concrete is such a widely used building material that some builders and homeowners believe that constructing good-quality concrete structures is easy. Actually, careful planning, workmanship, and use of quality materials is required if the concrete is to be problem-free.

Common Problems

Some of the more typical on-site problems, their causes, and measures to prevent them are covered here. Other problems may relate to the type of aggregate or other mix materials, and to soil conditions.

Crazing. A network of fine surface cracks. May collect dust and moisture. Not as serious as scaling. Does not necessarily seriously affect concrete quality.

Crazing and Dusting

Crazing (random surface cracks) and dusting each have a unique cause, but both also share a common cause—finishing when bleed water is on the surface.

Bullfloating, machine floating or hand floating, and troweling while bleed water is present can cause the sand and cement to work to the surface, mixing with the bleed water. The result is a sandy mix with a high water-cement ratio at the surface of the concrete. The result can be crazing or dusting, or even delaminating of the concrete surface. Sprinkling water or cement on the concrete surface to make finishing easier will have the same effect and should not be permitted.

Using too wet a mix contributes to bleeding, which can cause crazing or dusting. Poor curing worsens these problems.

Air entrainment, on the other hand, lessens bleeding and thereby helps prevent crazing and dusting.

Crazing can also be caused by low humidity, rapid surface drying, or too much cement in the mix. Excessive cement generates greater stress when the concrete attains final set; this is another reason that cement should not be used to absorb excess surface water during finishing.

Dusting. Fine dust that rubs off on finger. Is not necessarily a sign of serious deterioration.

Dusting is also sometimes caused by carbon monoxide resulting from use of unvented heaters. The carbon monoxide from use of such heaters in an enclosed space is also dangerous to workers and can cause death.

- Dry heat, even without exhaust gases, also causes dusting.
- A concrete mix that is rich in sand and low in cement can cause dusting.
- Certain chemical surface treatments help harden a surface damaged by dusting. If the dusting is only a surface problem, the surface of the concrete can be ground to expose stronger concrete.

Scaling

Scaling (flaking or peeling at the surface) can be caused by freeze-thaw cycles and by deicers. Scaling can happen even in enclosed areas such as garages if deicer lands on the concrete. Damage from freeze-thaw and deicers can also cause the concrete to spall (develop oval cavities at the surface). Therefore, concrete should be protected against freeze-thaw and deicers. (See "Scaling and Spalling, Freeze-Thaw, and Deicers," following in this chapter.)

Scaling. A serious problem. Surface scales off, leaving a roughened, irregular surface. May lead to other problems. May be a particular problem on driveways.

Good concrete should have a low water-cement ratio and other materials in the correct proportions; it should also be properly cured and air-dried so that it can resist scaling and spalling.

Measures to prevent scaling include—

- Use a reliable ready mix supplier. The supplier is responsible for the quality of the mix and its resistance to scaling. Unsound aggregates can cause popouts leading to scaling. The sand should be evenly graded to meet specifications; the concrete should be mixed thoroughly to ensure specified air entrainment; and the mix should have a low slump.
- Upon arrival at the jobsite, one addition of water is permitted to bring the load to the specified slump (but not in excess of the designated water-cement ratio). Thereafter, do not add water to increase the slump. Increasing the water-cement ratio will seriously affect the strength of the concrete and result in other undesirable properties.
- Make sure that good finishing practices are followed. The most important practice is not to work the concrete when bleed water is present. (See "Finishing the Concrete" in Chapter 4.)
- Finally, make sure that good curing practices are followed. Ponding is a good method. Allow enough air-drying time before deicers are applied. A period of 4 weeks of air-drying the concrete after curing is recommended.

Spalling

Spalling, as the term is used in this manual, denotes a deeper weakness in the concrete than surface scaling, although the two terms are sometimes used interchangeably. When concrete spalls, a large chunk breaks off, revealing a weakness that may extend to the full depth of a slab. The immediate cause of the spalling may be a blow to the surface or exposure of the concrete to freezing and thawing; or a chunk of concrete may break off at a slab edge or joint for no apparent reason.

Spalling. A deeper weakness than scaling with larger chunks breaking off.

The best way to prevent spalling—regardless of the immediate cause—is to use spall-resistant concrete. Spall-resistant concrete is also crack-resistant, scale-resistant, and watertight.

Follow these rules to reduce the likelihood of spalling—

- Use a good mix with a low water-cement ratio. If the proportions of any ingredient—sand, cement, or water—are incorrect, the concrete either will be weak or stresses will be created that lead to cracking and spalling. Soft, porous, and weak aggregates also can cause spalling.
- Use water-reducing admixtures to reduce the water content and still provide workable concrete.

- Air entrainment for concrete exposed to freeze-thaw or deicers helps to absorb freeze-thaw stresses.
- Improper finishing practices are more likely to result in surface imperfections such as scaling than to result in spalling, which is deeper into the concrete. Nevertheless, care should be taken not to finish concrete when bleed water is on the surface; also do not add water, cement, or sand to the surface to ease finishing.
- Proper curing is one of the most important factors for resistance to spalling. Good curing can greatly increase the strength of concrete—in fact, in some situations it may more than double the strength.

If edges and joints are poorly cured, they will be weak even if the rest of the concrete is properly cured. Moreover, the difference in curing between the edges and joints and the rest of the concrete may actually set up stresses that will cause cracks and spalling.

Other causes of spalling, such as corrosion of embedded metals and aggregate reactions, can be impeded by good curing.

Scaling and Spalling, Freeze-Thaw, and Deicers

Concrete is especially prone to scale or spall if it is exposed to ice and freeze-thaw cycles, particularly if deicer salts are used.

The following general guidelines can be used to make concrete more resistant to freeze-thaw and deicers—

- Make good, strong concrete. The water-cement ratio should not exceed 0.5 and the slump should be between 3 and 5 inches, depending on how it is to be placed. A water-reducing admixture allows less water to be used. Use at least 6 bags of cement per cubic yard of concrete and sound, well-graded aggregate. Place and finish properly.
- Air entrainment is essential for freeze-thaw and deicer resistance. The amount varies with air temperature and admixtures. The amount and quality should be monitored.
- Cure the concrete properly. Curing may need to be longer than ordinary in cool weather. If a normal Type I cement is used, 7 days of curing at 70 degrees F are recommended for extra protection. Type II high-early-strength cement sets quicker and cures more rapidly, so 5 days may be sufficient for curing.
- Air-dry the concrete. Air-entrained concrete should be air-dried for 4 weeks after curing to increase its resistance to deicers. In areas where freeze-thaw is a consideration, liquid-membrane cures applied in late fall may not allow sufficient time for air-drying before exposure to freeze-thaw or deicers, so another type of cure is recommended.
- Apply a surface treatment to help protect the concrete against deicers and lessen scaling and spalling, particularly if the air-drying period is cut short.

A breathable surface sealer of boiled linseed oil can be used as a surface treatment. The surface should be clean and dry before applying. Make two applications of the oil. Mix the first application with an equal part of solvent such as turpentine, naphtha, or mineral spirits. The second application can be pure linseed oil. Note: Linseed oil does not protect the concrete against freeze-thaw damage, only against the damage caused by deicers.

Surface treatments may need to be repeated in future years. Because oil treatments will produce a slippery surface until absorbed, it may be necessary to keep traffic off the concrete until sufficient drying has taken place.

Some authorities highly recommend a linseed oil emulsion both for curing and as a scaling-spalling preventative. This emulsion is not the same as linseed oil mixed with solvent.

Cracks

Concrete shrinks as it hardens. If it is restrained from moving, the resulting shrinkage stresses may be relieved by cracks.

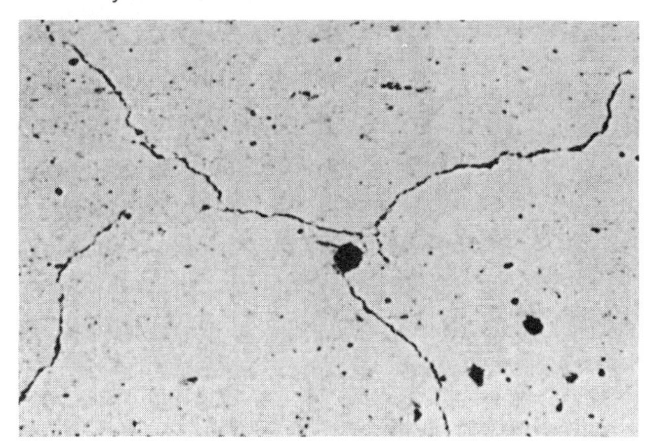

Cracking

The builder cannot hope to avoid cracks totally. However, controlling cracks is possible. The guidelines for controlling cracking are much the same as those for making concrete spall-resistant—

- Maintaining a proper water-cement ratio is the most important factor. Generally speaking, the more water that is used the weaker the concrete, with less strength to resist cracking.
- Good curing helps concrete gain strength before shrinkage stresses develop. With good curing the ultimate strength is higher and the ultimate stress is lower.
- Keeping the concrete temperature uniform during curing helps prevent cracks. Avoid sudden

temperature changes or abrupt termination of curing; make the cure uniform.
- The subgrade must be firm and uniformly compacted.
- Do not restrain the concrete. Use isolation joints at foundation walls and around columns. (See the sections "Controlling Cracking" in Chapter 1 and "Isolation Joints" in Chapter 5.)

Control where cracks may occur by—

- Making joints. To be effective, control joints should be one-fourth the depth of the slab (see the section "Locating Control Joints" in Chapter 5).
- Adding wire mesh or reinforcing steel bars to the concrete. The reinforcement reduces the size of the cracks, distributes them more uniformly, and keeps the aggregate interlocked at cracks to reduce displacement. (See "Installing Reinforcement" in Chapter 5.)

Plastic-Shrinkage Cracks

Plastic-shrinkage cracks usually form when the concrete is still plastic or not completely hardened. They are typically caused by rapid evaporation, when the water leaves the surface faster than it is replaced by bleeding.

Plastic-shrinkage cracks

Plastic-shrinkage cracks are more likely to occur in hot, dry, windy weather. However, they may occur in cool weather if evaporation is very rapid because of high winds, low humidity, or concentrated artificial heat. Very warm concrete in cold surrounding air may also cause evaporation and plastic-shrinkage cracks.

Plastic-shrinkage cracks can be prevented by using these techniques—

- Slow evaporation is recommended. Use windbreaks, temporary covers, or shades; cool the concrete or place it in the cool part of the day. Begin the cure immediately after finishing or even before finishing by covering the concrete with wet burlap, polyethylene film, or curing compound. (Sometimes a fine spray or a fog nozzle is used to keep the surface damp, but this is not recommended since water may build up on the surface. Water worked into the surface during finishing is a main cause of crazing, dusting, and scaling.)
- Hold water in the concrete in dispersion longer to decrease the moisture difference between the surface and the interior. Use air entrainment or increase the fineness of the sand.
- Polypropylene, nylon, polyester, or metal fibers added to the concrete mix can reduce the formation of plastic-shrinkage cracks by providing the fresh concrete with added surface tension and by reducing the migration of bleed water.

If plastic-shrinkage cracks begin to form during finishing, tamp the surface adjacent to the crack to bring the crack together and then refloat.

Blisters

Blisters range from ¼ inch to 4 inches in diameter and are about ⅛ inch deep. They appear when the concrete surface is sealed while air or water is still rising to the surface.

The following general rules help prevent blisters—

- Do not work the concrete too much. Overworking it may cause the aggregate to settle and bleed water to rise.
- Do not seal the surface too soon. Use a wood bullfloat on non-air-entrained concrete to avoid sealing the surface.

Use only magnesium or aluminum tools on air-entrained concrete. Air-entrained concrete is not as likely to blister since less bleed water is present.

Bugholes

Bugholes, sometimes called blow holes, are small cavities that appear because of air bubbles trapped in the surface of vertically formed concrete. They can be reduced by adjusting the sand content of the mix and by proper vibration, and corrected by sacking or rubbing.

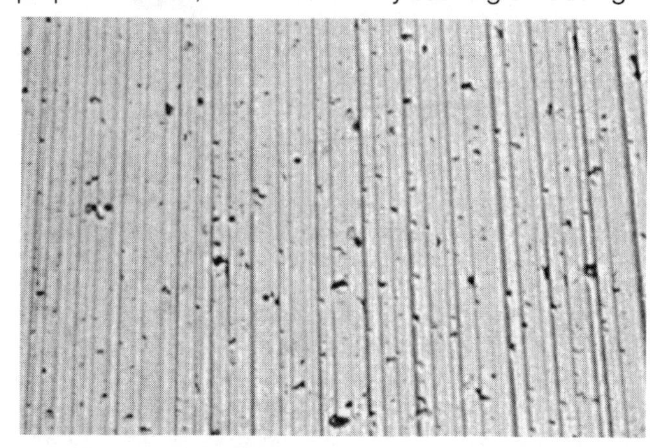

Bugholes, also called blow holes

Pitting

Small cavities, called pitting, develop from corrosion and disintegration of the concrete surface. Good curing can help prevent pitting. The damage can be corrected by sacking or rubbing.

Popouts

Popouts are shallow, conical sections that break away from the surface of concrete. They occur when aggregates split or expand near the surface of the concrete, creating internal pressure that causes a piece of the concrete to pop out. Use of good-quality aggregate will avoid popouts.

Popouts

Curling

Curling is when the slab rises slightly at the corners and edges. It occurs when the top of the slab shrinks more (or faster) than the bottom of the slab.

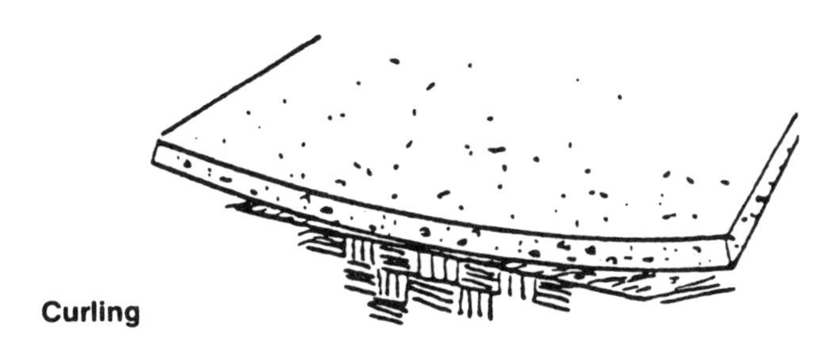

Curling

The most common reason for curling is too much variation in moisture or temperature between the top and the bottom of the slab. With good curing, concrete is less likely to curl because the moisture content and temperature of the concrete are kept uniform from top to bottom.

Other factors that may contribute to curling include slabs that are too thin and control joints that are too far apart.

The remedy for curling is water curing—for example, ponding—until the concrete flattens as much as possible. Then the concrete should be saw-cut into smaller squares.

Blowups

A blowup is the buckling upward of concrete on both sides of a joint or crack. Blowups usually occur in hot weather when the joints do not absorb the expanded concrete because they are too narrow, too far apart, or filled with incompressible material such as sand or gravel.

Blowup

The remedy for blowups is to provide adequate expansion joints at the intersection of moving slabs and fixed position elements. Concrete shrinkage cracks normally provide enough space to allow for future expansion of the concrete in hot temperatures. However, in order to allow these shrinkage cracks to flex, they must be kept clean of sand, dirt, and other debris. This means that to control cracks, sawcuts should be filled with a flexible joint sealer which must then be maintained in good condition for the life of the slab. (See "Sealing Control Joints" in Chapter 5.)

Efflorescence

Efflorescence is a crystalline deposit of soluble salts, usually white, that forms on the surface of masonry or concrete. While unsightly, it is generally harmless to the masonry or concrete.

Three conditions are needed for efflorescence to occur: (1) soluble salts must be present, (2) moisture must be present to pick up the salts, and (3) evaporation or hydrostatic pressure must cause the salt solution to move.

The best way to prevent efflorescence is to make good, watertight concrete with a low water-cement ratio, properly graded aggregate, the right amount of cement, air entrainment, well-consolidated placement, and good curing.

Vapor barriers on exterior walls and below the slab may help prevent efflorescence.

To remove efflorescence, dry-brush or lightly sandblast the concrete and then flush the surface with water.

If this treatment does not work, a 5- to 10-percent solution of muriatic acid may help. Wear gloves and goggles when working with acid. For integrally colored concrete, a more dilute mix of about 2 percent muriatic acid should be used to prevent etching that may affect appearance. Before using the acid, dampen the entire surface. Dampen small areas at a time, but dampen the entire surface. Flush with water afterward until the acid is diluted sufficiently so that it is no longer dangerous to touch or to the environment.

An alternative is to use a 1- to 3-percent solution of phosphoric or acetic acid.

Discoloration

Concrete may appear lighter, darker, mottled, or otherwise discolored in random spots. Though many factors cause discoloration, possibly the principal cause is lack of uniformity in the mix, particularly variations in the amount of water in the mix. If the water-cement ratio varies, the color of the concrete will vary.

Lack of uniformity in curing can also cause color variation resulting from uneven drying of the concrete.

A cure using polyethylene or curing paper may produce a greenhouse effect where moisture condenses under wrinkles and runs down, spotting the surface where the water puddles.

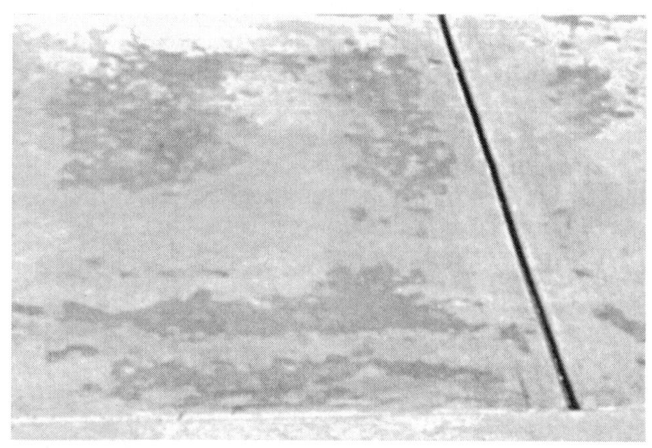

Discoloration

If the subgrade is not uniformly damp, the concrete can discolor.

To avoid discoloration and color variation, everything in the concrete mixture should be uniform. Do not order ready mixed concrete for the same job from different plants. Schedule the job so that the mix design, placing, finishing, and curing of the concrete do not vary.

Other causes of discoloration include calcium chloride added to the mix (especially if more than 2 percent calcium chloride by weight of cement is used) and hard-steel troweling for too long (particularly if calcium chloride is used). To remove discoloration from these sources, washdowns with water may be helpful. Stainless steel or plastic trowels are sometimes used on white concrete to help avoid discoloration.

Dusting wet concrete with cement to ease finishing can also cause discoloration—as well as crazing, dusting, and other surface problems.

Removing other discoloration such as ink, oil, urine, or tobacco requires special techniques.

Flash Set

Flash set occurs when the concrete sets rapidly before it can be finished. Flash set can be caused by adding hot water (140 degrees F or higher) directly to the concrete, using too much calcium chloride or other accelerators, or a combination of other factors. Flash set may happen in hot or cold weather.

Cold Joints

A cold joint is a weak bond that forms when one batch of concrete sets up before the next batch is placed against it.

To avoid cold joints, place each new batch against the preceding batch without allowing too much time between placements. Do not dump the concrete in separate piles.

Dished Surface or High Edges

A dished surface can result from finishing concrete that is low in the forms. Even if the concrete is struck off level with the forms, some settlement will occur, especially if too much water is used.

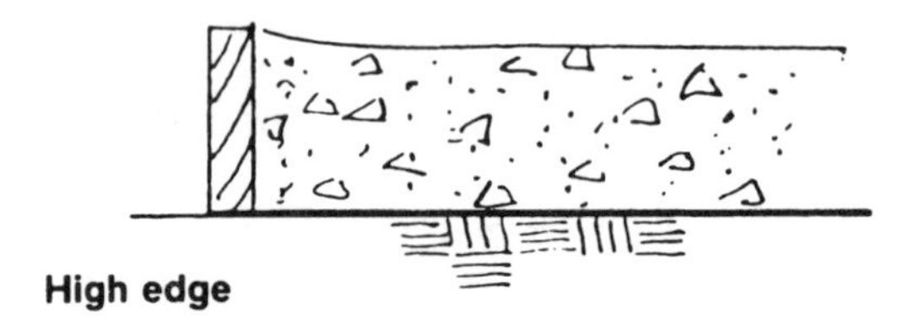

High edge

To prevent dished surfaces—

- Strike off the concrete a bit higher than the side forms to allow for settlement. Skilled and experienced cement finishers are able to do this by manipulating the strike-off straightedge.
- After floating, use the back of the trowel to straighten the edge and then finish the slab.
- Place a thin strip of wood on the top of the form, then remove it after the first floating and troweling. The strip will leave the concrete a bit high, letting the power float and trowel extend 4 to 6 inches past the edge. This helps keep the edge in a smooth, raised plane and prevents dishing.

Dished surfaces also may result from faulty bullfloating. As the mason pushes the bullfloat across the slab and then prepares to return, the bullfloat may dig in slightly. To prevent this, gently jiggle the bullfloat to loosen it before returning. Bullfloating a second time in a perpendicular direction, or darbying the surface in addition to bullfloating, may help.

Patching and Repairing

Regardless of whether the problem being addressed is a spall, a crack, or large-scale resurfacing and whether the patching material is concrete, epoxy, epoxy mortar, or latex concrete, there are two essentials for a good repair—

- The area to be repaired must be clean, rough, and sound.
- The patching material must be designed for minimal shrinkage.

Patching Spalled Areas with Concrete

The best approach to patching spalled concrete depends upon the size of the damaged area. For small- to medium-sized patches, saw or chisel the edges of the spalled area as shown; angle the cuts toward the undamaged concrete.

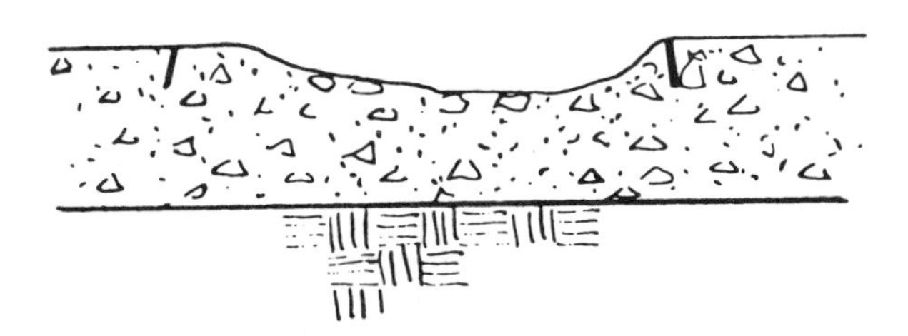

Chip out the area between. Leave the chipped area rough and make sure that the concrete is firm and sound.

Use an airblast or vacuum to clean the hole thoroughly. Then fill the hole with water and let it stand for a short time to soak into the rough surface.

Prepare a grout mix of 1 part portland cement, 1 to 2 parts fine sand, and enough water for a creamy mix.

Remove any water left in the hole after it has had time to soak in. The concrete should be damp but with no standing water.

Brush the grout over the entire area to be patched, including the sides of the hole.

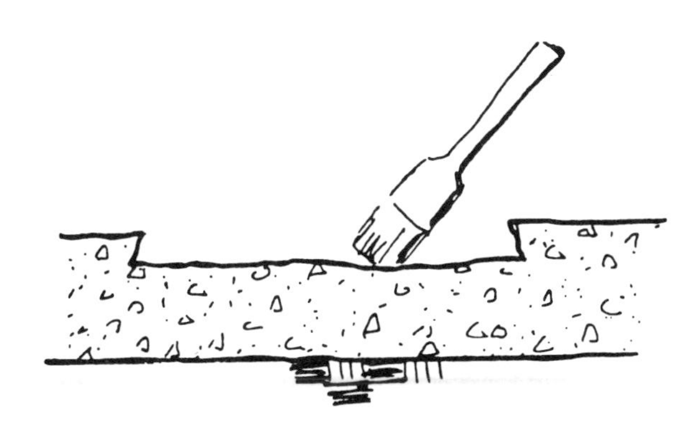

Make up a batch of concrete that is the same as the original concrete, except that—

- The slump should be nearly zero. That is, the mix should be very stiff and just wet enough to be worked into a ball by hand.
- The aggregate size should not exceed half of the patch depth. If possible, use concrete sand rather than mortar sand.
- Since the patch mix is likely to be darker than the surrounding concrete, white cement can be used in place of some grey portland cement. If time permits, experiment to determine how much white cement is needed. Typically, using about one-fourth white cement will produce good results.

After mixing, let the concrete stand for 15 to 30 minutes to minimize shrinkage before putting it in the hole.

Place the patch mix in the hole and compact it, overfilling slightly. In deep holes, pack the mix in 1-inch layers. Finish with a float and trowel. Cure for at least 3 days, or longer if possible. A water cure is recommended.

To make a good patch, remember that—

- Patch shrinkage can be a problem. Using zero-slump concrete modified with a high-range water reducer and a good cure will help.
- The surface should be roughened. Although acid is sometimes used, it can be dangerous. A scarifying machine is recommended instead. Sandblasting is all right provided it roughens the surface.
- The surface to be patched must be clean and free of dust, debris, oil, and other foreign matter.
- If the old concrete contains reinforcing bars and the patch is as deep as the bars, place the patching material about ¾ inch around the bars.
- If the patched area is over a joint, the patch must be jointed directly over the original joint.

When patching cracks, follow the same procedure used for patching spalled areas. The crack should be cut out in the same way or routed out and thoroughly cleaned.

Patching Spalled Areas with Epoxy

Epoxy products vary greatly; therefore, the manufacturer's recommendations should always be checked before using. The following comments are meant only to give a general idea of how to use typical epoxies. Before patching with epoxy, the surface to be patched should be—

- Sound, clean, and slightly roughened.
- Dry or basically dry with no water vapor moving up through the concrete. Special epoxies are available for damp concrete.
- At the proper temperature, usually between 60 and 100 degrees F. Special epoxies are available for temperatures below 40 degrees F and for the 40- to 60-degree F range. Heating the epoxy speeds up hardening. Leaving it in the mixing pot builds up heat and speeds hardening. At low temperatures the concrete may need to be heated to at least a 3-inch depth to accelerate hardening.

Sand—and sometimes a larger aggregate—is generally added to the epoxy before patching. Because epoxy and concrete have different expansion-contraction factors, the addition of aggregate helps make the epoxy's expansion and contraction more like that of concrete. If aggregate is not used, the epoxy patch tends to break loose from the concrete. Only sand is added to thin patches. Sand and coarse aggregate are added to thicker patches, but the aggregate size should not exceed one-third the thickness of the patch.

Remember that different epoxies are used for different situations such as indoor or outdoor patching, high or low bonding temperatures, and patching new concrete to old.

Making an Epoxy-Mortar Patch

When patching with epoxy, make vertical or sloping cuts at the spall edges as necessary to avoid a feathered edge. Slightly roughen the area to be patched. Chipping and sandblasting the surface are recommended. Clean the concrete surface of all oil, grease, and loose concrete.

Dry the surface unless a special epoxy is to be used. Be sure no moisture is rising through the concrete.

The concrete and air should be in the right temperature range, usually 60 to 100 degrees F unless a special epoxy is used. The concrete should be heated if necessary.

Prime the roughened and cleaned concrete with epoxy. If the spall has steep sides or reinforcing steel, the mix might run out of the patch area. Check the manufacturer's recommendations before using.

Add aggregate to the epoxy to make epoxy mortar. Use sand alone for patches that are no deeper than ¾ inch—that is, concrete sand with a range of particle sizes rather than mortar sand with more uniform particles. Use coarse aggregate along with the sand for patches deeper than ¾ inch. The aggregate should be no larger than one-third the patch depth. The epoxy binder-aggregate ratio by weight is generally in the 1-to-7 range, depending on the maximum size of the aggregate.

Apply the epoxy mortar to the patch area. The mortar should be applied before the epoxy primer hardens to avoid having to sandblast the primer before adding mortar. If the patch is more than 5 inches deep, build it up in two or more layers to prevent low spots and reduce heat buildup and subsequent contraction. Stir the mortar to release any entrapped air which would reduce the epoxy's strength and increase the thermal expansion.

Finish the patch with a trowel.

Finally, broadcast a light layer of sand over the patch and the surrounding primed area until it has a uniform appearance.

When patching a spalled area with epoxy mortar, be sure the joint is clean and roughened. A steel plate or trowel can be used as a backstop. Remove the backstop after striking off the mortar.

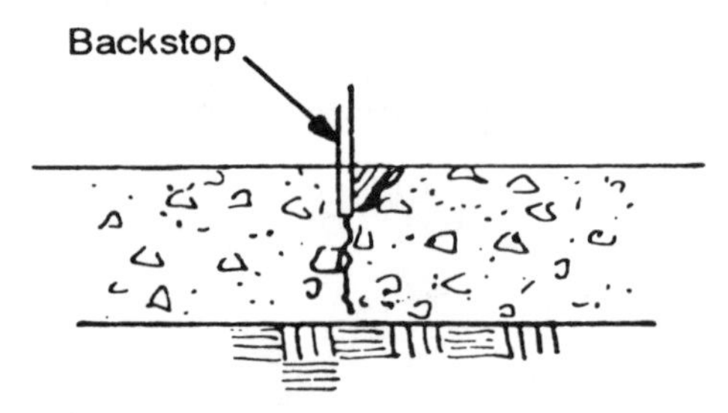

Additional Uses for Epoxy

- Epoxy coatings with mineral particles can be used for skid-resistant surfaces.
- Epoxy can be pumped into cracks to seal them, but check the manufacturer's recommendations first.
- Concrete can be coated with epoxy for greater resistance to deicers, moisture, and wear.
- Epoxies can be used to bond fresh concrete to old concrete, or old concrete to old concrete if a chunk breaks off (see manufacturer's recommendations). A special epoxy is needed to bond fresh concrete to old, and in some cases may be needed to bond old concrete to old concrete (for example, in bonding horizontal surfaces).

When using epoxy, be sure to follow the manufacturer's recommendations for safety and for cleaning tools.

Patching Spalled Areas with Latex Mortar

Latex added to concrete improves ductility, durability, adhesive properties, and tensile and compressive strength. This type of concrete is used primarily for patching and overlays.

As with the epoxies, different latex modifiers are used for different conditions. Check manufacturer's recommendations before using latex products.

Latex modifiers in ordinary mortar provide good adhesion, resilience, elasticity, and durability. They can be used for both interior and exterior patching.

Latex can be applied to the base concrete, and then ordinary concrete patching mortar can be used for the patch. Latex also can be mixed with the water for the patching concrete, making a latex mortar.

Some general notes on using latex—

- Latex-modified mortar can turn a green color during the initial cure; but if a good-quality latex modifier is used, the concrete will be a normal grey color after it is cured.
- Latex mortar has a high slump of up to 8 or 10 inches.
- Latex mortar should not be air-entrained because it loses stability with air entrainment.
- Latex mortar should be applied on a damp base and must be struck off and finished rapidly because of its short working life.
- Since latex mortar clings tenaciously to tools, rapid cleanup of equipment is essential.

Other Patching Agents

Various resins, resinous emulsions, acrylics, and polymers are also used as bonding agents for concrete patches. Characteristics vary with the product and the manufacturer.

Resurfacing Slabs

A topping may be either bonded to the base slab or left unbonded. The topping and base slab should either operate as a unit or be totally independent. The choice of whether or not to bond influences how the topping is constructed, including the control joints.

Do not bond the topping if the base slab is basically sound. Topping depth should be a minimum of ¾ inch for bonded slabs and a minimum of 3 inches for unbonded slab.

Applying a Bonded Topping

Follow the general rules given for patching. Surface preparation is extremely important: the concrete should be clean and roughened. Use a low water-cement ratio to give low shrinkage, high strength, and a good bond.

Roughen the surface with a scarifying machine, a jackhammer, or a chipping or rough-grinding tool. The surface does not need to be extremely rough, but should be slightly roughened all over. If the roughening operation goes beyond a reinforcing bar, roughen a minimum of ¾ inch around the bar.

Clean the surface with a vacuum or its equivalent, wash with water, brush, and reinspect. If acid is used to roughen the surface, use litmus paper to ensure that all the acid has been flushed away.

Dampen the concrete surface before the grout is brushed into place, particularly if the temperature is 60 degrees F or hotter. Ideally the base concrete should be dampened several hours ahead of time. But there should be no standing water when the grout is placed.

Apply a 1⁄16- to ⅛-inch layer of bonding grout just before the resurfacing concrete is placed. This grout should be 1 part portland cement, 1 part fine concrete sand, and ½ part water. Mix the grout to a thick, creamy consistency and broom it onto the surface. In no case should the grout dry before the topping is placed.

The concrete topping should contain pea gravel, crushed stone, or other suitable coarse aggregate graded up to a top size of ⅜ inch and well-graded concrete sand. The aggregate should be no larger than half the topping thickness. A mix of 1 part portland cement, 1 part sand, and 1 ½ to 2 parts coarse material is recommended. Use as little mixing water as possible. The water-cement ratio should be no more than 0.44, or 5 gallons of water per 94-pound bag of cement, including any free water in the sand. The slump should be near zero. Special equipment such as vibratory screeds, tampers, or power disk-floats may be needed to place and compact the concrete.

This type of low-slump topping probably should be mixed on the job. Ready mix companies may have difficulty controlling such a mix.

- Strike off about ⅜ inch high.
- Finish the concrete by power floating because this compacts such a dry mix better than hand floating.
- Control joints for a bonded topping slab should be exactly over the joints in the base slab; they should be at least as wide as the base joints, and extend completely through the topping.

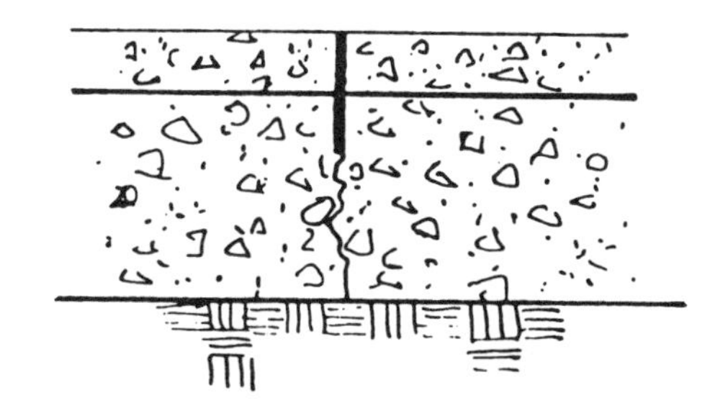

Joints can be formed by inserting strips in the base joints to serve as fillers for the topping-slab joints; or the joints can be sawed out and filled with a sealer.

Curing is even more important in bonded resurfacing than in ordinary concrete work, particularly in hot, dry, or windy weather. Good curing is essential not only because of the thinness of the topping slab, but also to control shrinkage-stress and to promote a good bond with the base slab.

Wet curing by ponding is recommended.

When curing a bonded topping, the following recommendations apply—

- Start as soon as possible.
- Cure for 7 days or more in temperatures from 50 to 70 degrees F.
- Cure for at least 5 days in hot weather—70 degrees F or higher—or when using high-early-strength concrete.

Applying an Unbonded Topping

The minimum recommended thickness for an unbonded topping is 3 inches. The base course should be covered with plastic sheeting or felt that is as wrinkle-free as possible. (Some contractors maintain that sand over felt helps prevent curling.) Good curing and shorter control-joint spacing should also reduce the curling effects.

Approximately 30 pounds per 100 square feet of wire mesh is needed to control shrinkage cracking. Place the mesh mid-depth in the topping and cut it at each joint. The maximum aggregate size must be no more than one-third the thickness of the unbonded topping.

Maximum slump should be 1 inch for areas of heavy usage. Since concrete this stiff is hard to place and finish, high-range water reducers can be used to increase the workability of the mix without increasing the amount of water. Added water reduces concrete strength and increases shrinkage.

Joints should be one-half the topping depth.

Power floating and troweling are usually required, but the final hard-troweling should be done by hand.

Correcting Bugholes and Form Marks

Bugholes and small projections caused by form marks can be corrected as follows—

Moisten the concrete area to be worked. Cover the concrete with a cement and sand mortar, then rub it over the entire surface with a wood float or medium-grade carborundum stone. A steel float would darken such a surface and make the corrected area obvious.

The mortar is made of 1 part portland cement to 1 ½ parts concrete sand. This is similar to a grout dressing (see "Grout Dressing" in Chapter 4). For a better appearance, cover an area wider than the defective area.

Some excellent proprietary cementitious products are available for eliminating bugholes with much less labor. Builders should consult with their dealers.

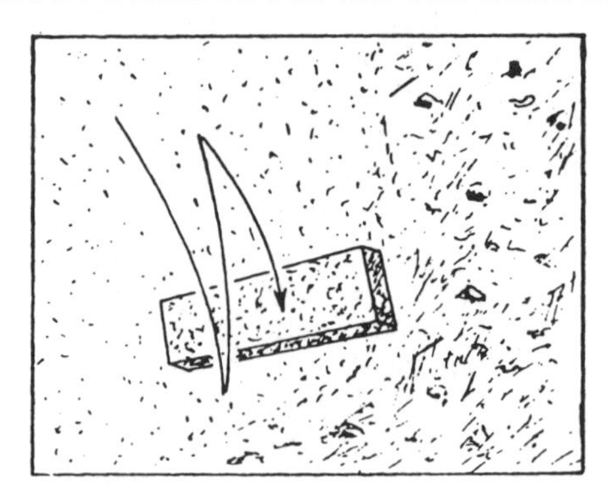

Carborundum stone

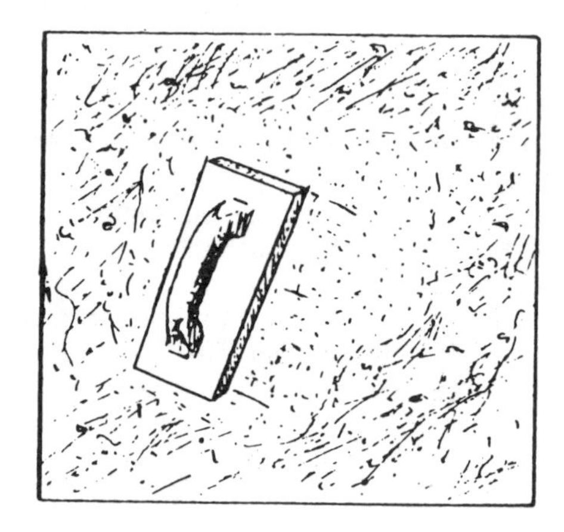

Wood float

Patching Tieholes

Wires are often used to tie the two sides of a wall form together. When these ties are cut or broken off to remove the forms, a hole in the surface, called a tiehole, remains. To patch a tiehole, clean the hole, wet it, and apply a cement grout of portland cement and water. Then, using a short rod, ram earth-dry mortar into the hole until it is completely filled.

If a form tie-wire has been left exposed, the surrounding concrete must be chiseled away so that the wire can be cut far enough down to avoid rust stains—1 ½ inches for concrete exposed to weather.

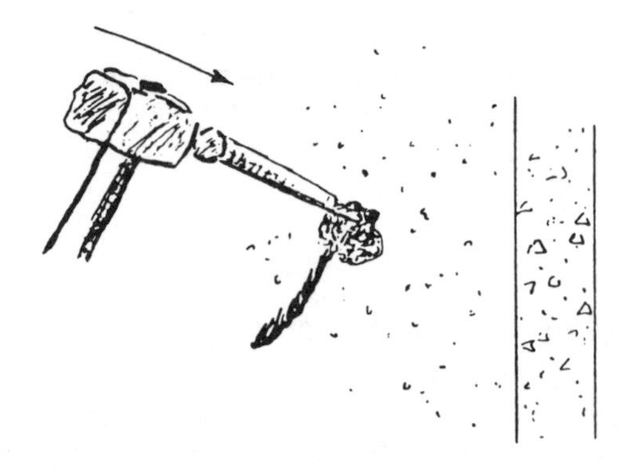

To patch this type of hole, use a mortar that is about the same as that in the original concrete. For example, if the original concrete consisted of 1 part cement to 2 parts sand and 4 parts large aggregate, make a mortar patch of 1 part cement to 2 parts sand.

Since large patches tend to dry dark, a bit of white cement should be used in place of some of the grey cement. About 20 percent white cement is typical. Experiment with proportions and let the materials dry out for a true comparison. If possible, use concrete sand since it has a range of sand sizes.

Repairing a Broken Corner

A broken corner that is still in good shape can be glued back in place. Clean the area and mix an epoxy. Butter the broken piece with the epoxy, hold or brace it in place for about 10 to 15 minutes, then clean off the excess epoxy. If the broken-off corner is missing or not in good shape, the corner can be repaired as a spalled area (see "Patching Spalled Areas with Concrete" and "Patching Spalled Areas with Epoxy" in this chapter.)

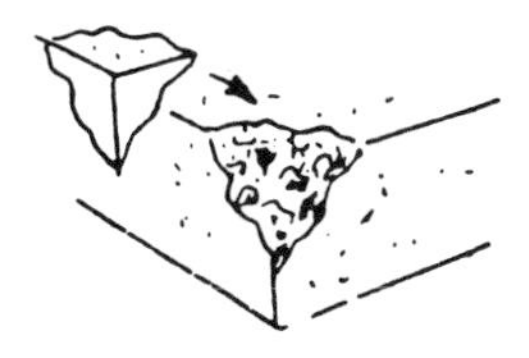

Repairing Honeycomb

Honeycomb repair is similar to patching spalled areas. The honeycombed area should be chipped out and the edges cut away as if for spall patching. Clean and wet the area and brush on a bonding grout containing cement and water, but no sand.

The area is then immediately filled with a mortar of earth-dry consistency. A steel trowel may be used for filling, but a wood float should be used for finishing.

If needed, form a board-mark across the completed patch by lining up a board with the existing marks and rubbing it back and forth.

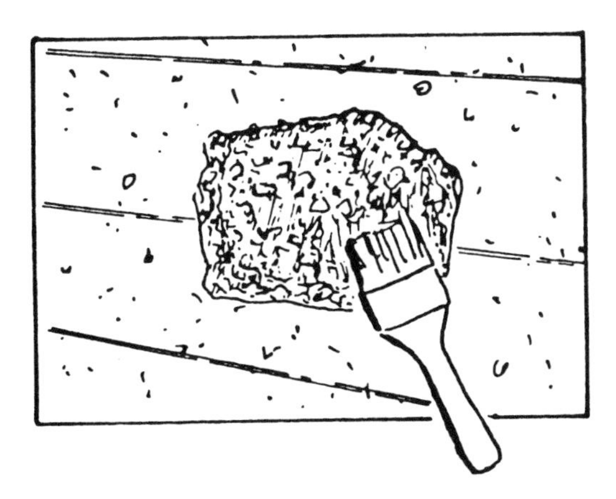

Cure the patch for at least 3 days in temperate weather and longer in cold weather.

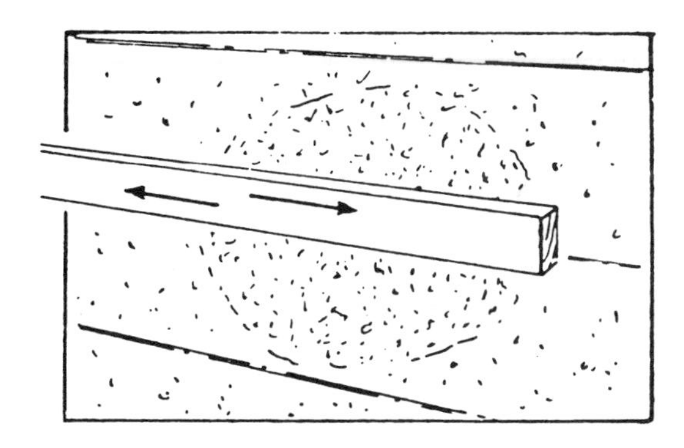

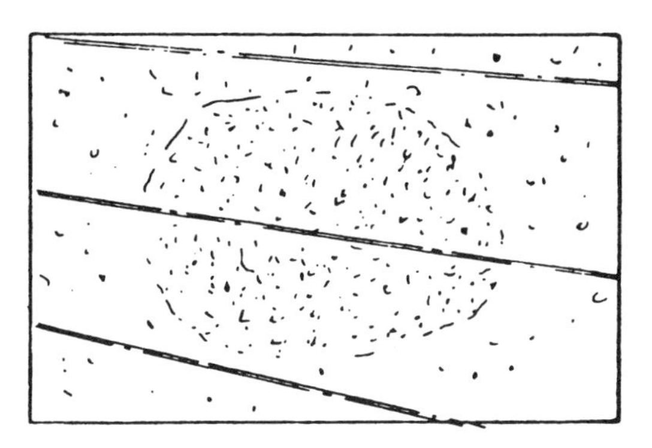

Making Concrete Watertight

Generally, the same things that make concrete strong and crack resistant make it watertight, including—

- Low water-cement ratio: 0.50 by weight.
- Good cement proportions: 6 bags per cubic yard of concrete for 1-inch maximum aggregate.
- Entrained air: 6 ± 1 percent for 1-inch aggregate.
- Nonporous aggregate.
- Handling concrete in a way that avoids segregation.
- Good curing: 7 to 14 days, or longer if needed.

Watertightness should not be misunderstood to mean that the passage of water vapor is prevented. The best-cured concrete will still allow the passage of some water vapor.

Certain admixtures, called dampproofing or permeability-reducing agents, may reduce water passage in concretes that have a high water-cement ratio, low cement content, or excessive fine sand. However, some concrete experts maintain that some of these admixtures may actually increase water transmission in well-proportioned mixes, since more mixing water may be needed.

Good site drainage is a very important part of your overall waterproofing measures. Finish grading should include a slope to draw water away from the foundation of the house.

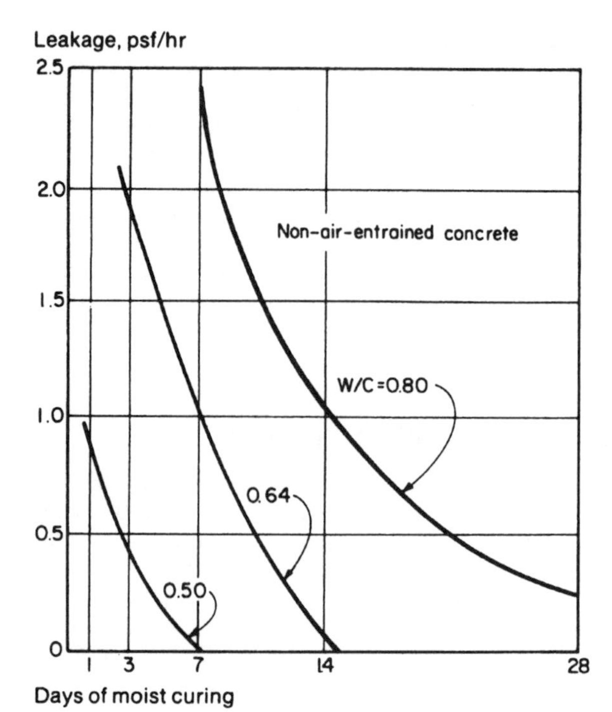

How water-cement ratio (w/c) and curing affect watertightness.

Special Waterproofing Measures

Even if the basic concrete is sound, special waterproofing measures may be needed—especially if the site is unusually wet. The following measures help control moisture in special cases.

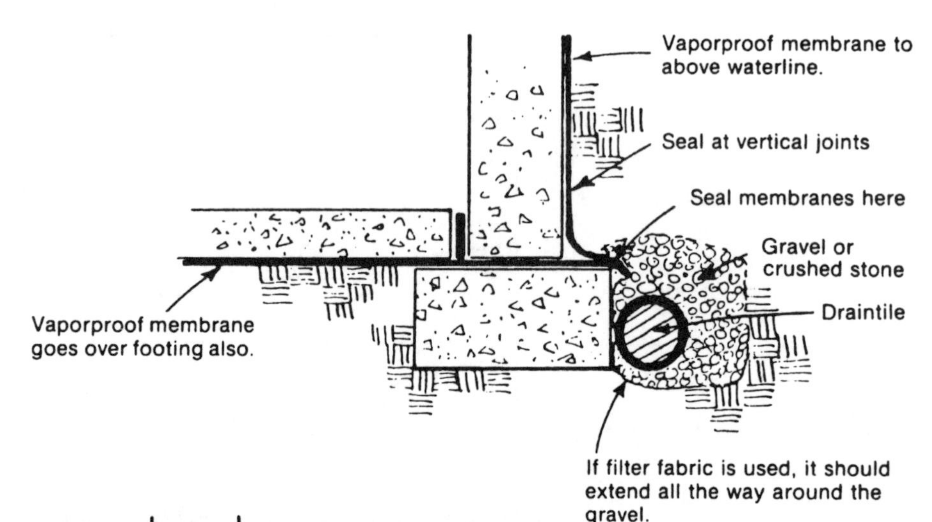

An alternate method for waterproofing when the water table is not above the slab and the inside wall is to be covered.

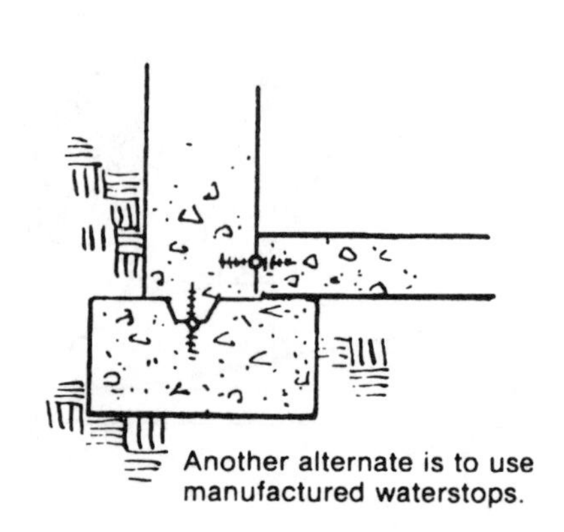

Another alternate is to use manufactured waterstops.

Waterproofing a Slab

A vapor barrier to waterproof a concrete slab can be an asphaltic curing compound or a sheet material such as polyethylene film, rubber, or asphalt. The base slab usually cures well because of the barrier, thereby gaining strength and watertightness. This system can be costly, however, because it requires placement of two slabs rather than one. Builders using a vapor barrier should be sure to check code requirements affecting their placement.

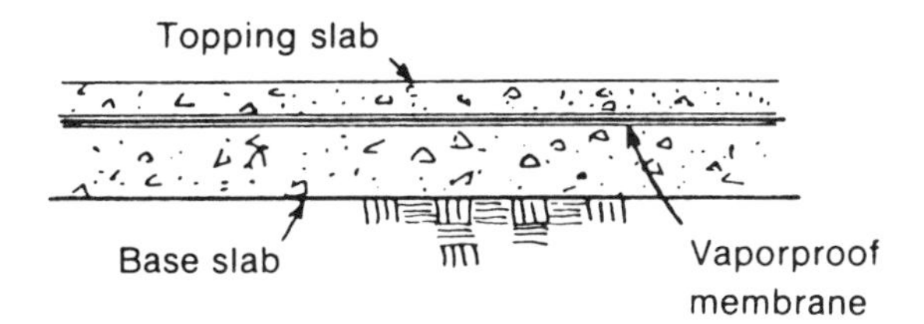

VAPOR BARRIER RATINGS

Material	Perm rating
15-pound roofing felt	0.6 to 2.0
6-mil polyethylene film	0.06
55-pound roofing felt	0.03 to 0.08
Butyl rubber sheeting	0.002
⅛-inch asphalt panels	0

NOTE: *Perm* refers to vapor transmission. The lower the perm rating, the better the barrier.

Roofing Felts

Roofing felts provide a relatively inexpensive barrier, but they are less effective than other materials and do not last long. Sections should be lapped 4 to 6 inches and sealed with hot asphalt. A 55-pound felt is more effective than a 15-pound felt but is not much more durable.

Polyethylene Film

Polyethylene film is another relatively low-cost waterproofing method, but it is easily punctured during installation, which defeats the purpose of the barrier. Wide sheets should be used and lap joints avoided. If lap joints must be used, they should be at least 6 inches wide. Sometimes builders spread sand over the gravel or crushed-stone fill to protect the film against puncturing. Polyethylene-coated kraft paper is more durable and has a comparable permeability rating.

Butyl Rubber Sheeting

Butyl sheeting is a good, rugged, long-lasting barrier. It is expensive, however, and may need to be prefabricated for the job. Splices are sealed with an adhesive.

Asphalt Paneling

Asphalt paneling is very effective, rugged, and less likely to be damaged by construction work. The panels are ⅛-inch thick and come in 4x8 sheets. They are installed either by lapping or butt-jointing with 8-inch-wide strips overlaying the joints. The joints or strips are sealed with hot asphalt or asphalt mastic.

Waterproofing Wall Joints

Control joints in concrete walls may be at openings, within 10 feet of corners, and a maximum of 20 feet apart otherwise. Making basements watertight and waterproofing in flood zones both require special attention to details that are beyond the scope of this book.

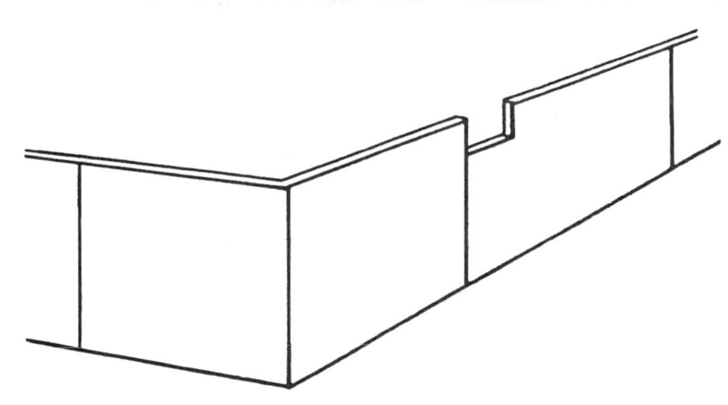

A ¾-inch wood strip beveled for easy removal is fine for 6-inch-thick walls, but total reduction of wall section must equal ¼ the wall thickness to make the control joint work.

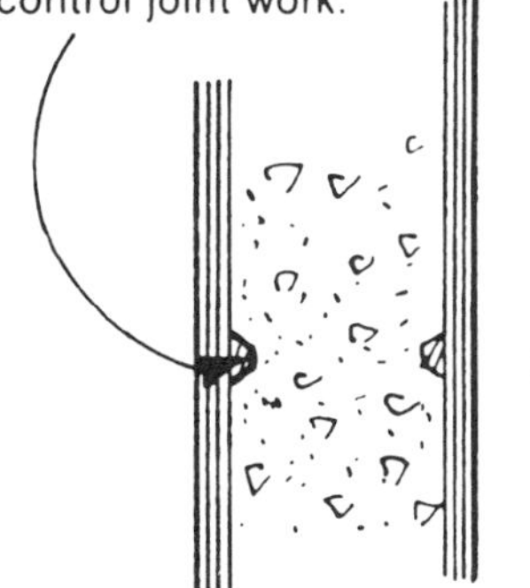

Forming the joint

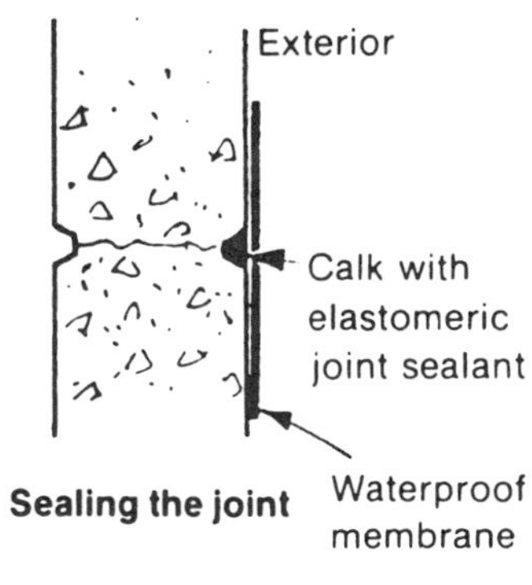

Sealing the joint

Summary of Concrete Problems

Some of the most common problems with concrete relate primarily to irregularities in finishing and curing. Other problems may stem from materials in the mix or even from soil conditions. The following points recap some preventive and repair techniques you can employ. To reduce the likelihood of many common problems—

- Prepare the subgrade carefully.
- Use spall-resistant concrete with a low water-cement ratio, and correct proportions of all ingredients.
- To protect the concrete against damage from deicers and freeze-thaw conditions, use air-entrained concrete, air-dry the concrete, and apply a surface treatment such as boiled linseed oil.
- Do not finish when bleed water remains on the surface, and avoid overworking the surface during finishing.
- Ensure uniform curing, taking particular care that edges and joints are properly cured.
- Space control joints carefully, and ensure that their depth is proportional to the depth of the slab.
- Use reinforcing fibers to reduce the formation of plastic-shrinkage cracks.
- Apply appropriate patching and resurfacing techniques if needed to correct minor problems.

Chapter 7

Special Finishes

Pleasing decorative finishes can be achieved with concrete construction. Variations of concrete surfaces are limited only by the imagination of the designer and the skill of the craftsman.

Colored Concrete

Four Ways to Color a Slab

Following are four ways to color a slab.

Dry-shake Method. Dry-shake is economical and produces bright color because the shake is not overly diluted by surface water.

One-course Method. One-course coloring is also called integral color. The concrete is colored all the way through. This method is not widely used except for small areas because of the cost.

Two-course Monolithic Method. Colored concrete topping 1 inch thick is placed on a grey base concrete. The slab is monolithic; that is, the topping is cast so soon after the base slab is cast that the topping and the base become one unit. Do not place the topping on bleed water.

Two-course Bonded Method. The bonded method is similar to the monolithic method except that the colored topping is bonded to the base slab at some later time. The base course is placed to within 1 inch of grade and the surface left rough for a good bond. The base should be high-quality concrete that is thoroughly cured and kept clean. If a liquid membrane is used for curing the base, it may need to be stripped before the colored topping is applied. Since stripping may be difficult because of the roughened surface, other curing methods are preferred.

After construction activity has slackened—possibly some weeks after the base course is placed—the colored topping is placed. Clean the surface of the base slab first. Thoroughly wet-cure the topping.

Making Good Colored Concrete

Use minimal pigment. Generally using less of a strong pigment is better than more of a weak pigment, and using less of a dark color is better than using more of a light color.

White portland cement produces brighter colors, especially when it is used with a light-colored sand. Some authorities recommend white cement even for dark colors.

The only way to maintain uniform color is to make sure that everything done to the concrete is done uniformly. The materials must be exactly proportioned each time. The color pigment must be the same—if possible, it all should be bought at the same time and place. Every batch of concrete should have the same mixing time, and the mixer should be kept very clean. The tools must be kept immaculate. The slump should be the same each time—about 4 inches. The subgrade should be uniform in dampness and support. The thickness should be the same. The finishing operations should be identical; extra workers may be needed to assure the same finishing operations each time. Finally, the curing should be the same and should be a type that does not leave streaks.

Cure colored concrete by ponding. This is recommended because polyethylene cures may cause a greenhouse effect and leave spots or blotches. Whatever cure is used, it should be the same for the entire job to keep the color uniform. A 14-day cure at moderate temperature is recommended. A special wax-cure compound applied by roller is available for colored slabs. It may even be available in the same color as the slab. If a wax-cure compound is used, the surface should also be protected with paper or some other scuff-resistant material.

Use air-entrained concrete to minimize bleeding, which can cause streaking or variations in color. Some color pigments may reduce air entrainment slightly.

Generally, calcium chloride should not be used for colored concrete. When used in cold weather, chlorides such as calcium chloride may leave efflorescence on the surface, particularly if more than the recommended amount is used. Other admixtures containing chlorides may also affect the surface.

Use the right type of trowel. Some experts recommend a plastic or stainless-steel trowel for colored concrete. Regular steel may rub off and leave stains.

Watch for a fast set. Colored concrete may set faster than regular concrete. Troweling too late can discolor the concrete and overtroweling can cause burns or darkened spots.

White Concrete

Cleanliness is most important when working with white concrete. Use special bins to store the cement and aggregate, special equipment for mixing (being especially careful to prevent contamination from iron particles), and clean forming equipment and form-release agents. Avoid stains from steel after the concrete is laid.

Mixing, Handling, and Finishing

Waterproofing helps maintain whiteness. If the concrete is to be wet often, either from exposure or cleaning, a water-repellent admixture or a waterproof type of white cement is sometimes used to help the concrete retain its whiteness.

The water-cement ratio should be low. Use no more than 5 gallons of water per bag of cement (0.44 by weight). Include the moisture content in the sand when calculating water quantity.

Mix proportions are important. For a ⅜-inch topping course for foot traffic only, the recommended mix proportions are 1 part white cement to 2 ½ parts white sand. Such a thin topping should be placed monolithically—that is, very soon after the base slab is placed so that the two layers become one unit.

Mix proportions for a heavy-duty topping should be 1 part white cement, 1 part white sand, and 1 ½ to 2 parts pea gravel or crushed stone graded percentage by weight as follows—

Passing ½-inch sieve	100%
Passing ⅜-inch sieve	95-100%
Passing No. 4 sieve	40-60%
Passing No. 8 sieve	0-5%

Spread the topping with care because it may discolor if materials in contact with the base concrete are pulled up.

Mechanical floating is preferred, since a stiffer mix can be used.

Dry-Shake Technique

Dry-shake is a method of finishing concrete to change its color or to increase its hardness, wear, or slip resistance. Some dry-shakes can be used to add sparkle or glitter to the concrete surface.

The technique for applying dry-shakes is similar, whatever its purpose. Materials are shaken onto the surface and worked into the concrete.

Applying Dry-Shakes

Strike off and bullfloat the base concrete. Wait until the surface is ready for floating. It should not have any freestanding water and the weight of an adult should make no more than a ¼-inch impression in the concrete. Float the surface by hand or power. Floating helps prepare the surface for the shake by bringing moisture up and removing ridges or depressions. (All tooled edges and joints should be run before and after applying the dry-shake.)

Next, spread the premixed shake material by hand evenly over the surface. Use about two-thirds of the material for this operation. Bend low over the slab and let it sift through the fingers. Shake and float the edges first since they tend to set up first. If the shake is applied too early, it will sink into the slab. If it is applied too late, not enough paste remains to finish the concrete correctly.

When the shake material is uniformly damp, float it into the surface. A power float is preferred.

Immediately spread the remaining shake material at right angles to the first application.

Then float the surface again after the shake has absorbed moisture.

The concrete may be troweled after the final floating. Flat-trowel first and then use a smaller trowel with an increased angle after the concrete hardens.

Hardness and Wear and Slip Resistance

Hardness and wear resistance come primarily from the concrete. If the water-cement ratio is low, the mix is right, and the finishing and curing are done properly, the concrete will be much harder and more wear-resistant.

Special aggregates that increase hardness and toughness are available. Special liquid coatings such as epoxies can provide curing, hardening, sealing, chemical resistance, dustproofing, antispalling, and nonslip surfaces.

Wear- and slip-resistant aggregates can be added to the concrete either in special toppings or as dry-shakes. They can provide decorative effects as well. (For placement of special toppings, see the manufacturer's directions and "Resurfacing Slabs" in Chapter 6.)

These shakes can be traprock, granite, quartz, emery, corundum, silicon carbide, aluminum oxide, or malleable iron. These materials are hard and must be worked into the floor with a float and pounded a bit if necessary until they are barely covered with paste. They may need to be dampened and coated with cement before floating them into the surface. The typical ratio by weight for coating with cement is 1 part cement to 2 parts aggregate. Further floating and hard-steel troweling will compact the material into the surface.

Two popular materials for achieving slip resistance are silicon carbide and aluminum oxide. They are extremely hard and sharp. Finish as if for a color dry-shake, and brush with a hairbrush several times after the shake is worked into the surface. Allow some time between brushings to make the particles glitter. Use ¼ to ½ pound of shake per square foot of slab.

An iron-aggregate mix or dry-shake should not be applied to a floor if calcium chloride is used. In addition, some silicon-base hardeners may obscure the color.

Exposed-Aggregate Finish

One method for creating an exposed-aggregate finish involves washing away a portion of the concrete from embedded rock. To use this method—

- Order gap-graded or special-color, large aggregate in your concrete mix.
- Specify and place air-entrained concrete with a low slump.
- Knock the rocks down and bring the cement paste up by using a strike-off and float, taking care to not depress the large aggregate too deep.
- Spray the surface with a surface retarder.
- Allow the concrete to set until the surface can be washed.
- Use a high-pressure washer or special nozzle to wash the paste away from the rock, about one-third down (two-thirds of the rock should remain embedded in paste).

Exposed-Aggregate Finish by Seeding

For an exposed-aggregate finish to be done by seeding, the base concrete should have a maximum 3-inch slump. Place, strike off, and bullfloat or darby in the usual manner; but keep the level of the surface slightly lower than the top of the forms to accommodate the extra aggregate. The forms should not be too high, however, because the surface must have good drainage when it is washed.

Seed wet, clean aggregate uniformly by shovel and by hand so that the entire surface is completely covered with one layer of stone.

Embed the aggregate initially by tapping with a wood hand float, a straightedge, or a darby.

For final embedding, use a bullfloat or hand float until the appearance of the surface is similar to the surface of a normal slab after floating. Do not overfloat or depress the aggregate too much.

The timing of the aggregate exposure operation is critical. In general, wait until the slab can bear a person's weight on kneeboards without making an indentation. Brush the slab lightly with a stiff nylon-bristle broom to remove excess mortar.

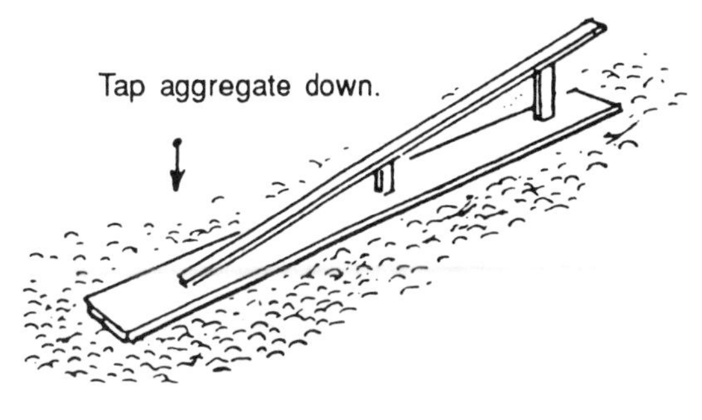

Hand float to smooth the embedded aggregate.

Next, dampen the brush and then brush and flush with water several times. Special exposed-aggregate brooms with water jets are available. If any aggregate is dislodged, delay the operation. Continue until the flush water runs clear and no noticeable cement film is left.

Brush to remove excess mortar.

Brush and flush with water to expose the aggregate.

Fine-spray the surface during brooming. Special exposed-aggregate brooms are available.

Continue washing and brushing until flush water runs clear and there is no noticeable cement film left on the aggregate.

A surface retarder can be sprayed or brushed onto the slab shortly after floating to permit a time lapse of several hours before the hosing operation.

Exposing operations should begin as soon as possible after danger of dislodging or overexposing the aggregate has passed. If moving about on the surface of the concrete is necessary, use kneeboards. Do not slide or twist the kneeboards; bring them gently into contact with the surface. If possible, stay off the surface entirely to avoid breaking the aggregate bond.

Typically, tooled control joints or edgings are not practical in exposed-aggregate concrete since the aggregate covers the entire surface. Joints can be made by sawing or using permanent strips of wood such as heart redwood.

Exposed-aggregate slabs should be cured thoroughly by using a method that will not stain the surface.

If a sealer is used, make sure it is compatible with the curing compound. Wait a few weeks after finishing before applying the sealer. Clean the concrete with an acid wash beforehand if necessary. Remove the acid, dry, and then seal.

Alternate methods for placing exposed aggregate are to use a bonded topping that contains the special aggregate or to use the integral method in which the select aggregate is contained throughout the slab.

Nonslip Finishes

Floating or troweling is the easiest way to obtain a nonslip surface, but this type of surface may also be the least wear resistant.

The mason can simply eliminate troweling after floating and leave a textured, floated surface, or swirl-float to create a pattern. Similar techniques include the popular broom finish, burlap drag, or wire combing.

Floated Swirl

To produce a swirl texture, the concrete is struck off, bullfloated or darbied, and a hand float is worked flat on the surface in a semicircular or fanlike motion, using pressure. Coarse textures are produced by wood floats, and medium textures by aluminum, magnesium, or canvas-resin floats. A fine-textured swirl is obtained with a steel trowel. Cork or rubber floats may also be used. Care must be taken to allow the concrete to set sufficiently so that these textures are not marred during curing.

Swirl finish, floated

Burlap Drag

Burlap makes a gritty surface of varying depths, similar to a wood-float finish. The depth of the finish depends on how soon after the finish the burlap is used. Some burlap has metal projections on the trailing edge to give deeper grooves.

Drag the burlap perpendicular to the flow of traffic.

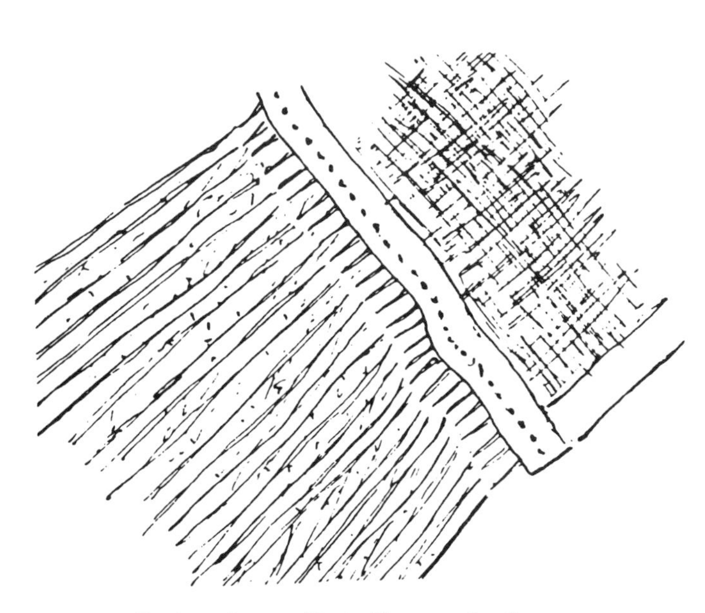

Burlap drag with trailing projections.

Wire Combing

Wire combing is done with a special tool designed for the purpose. The wire comb should be pulled across the concrete perpendicular to the traffic flow.

Transverse grooving with a steel-pronged rake gives an effect similar to wire combing; however, it requires more skill and is not recommended for large-scale use.

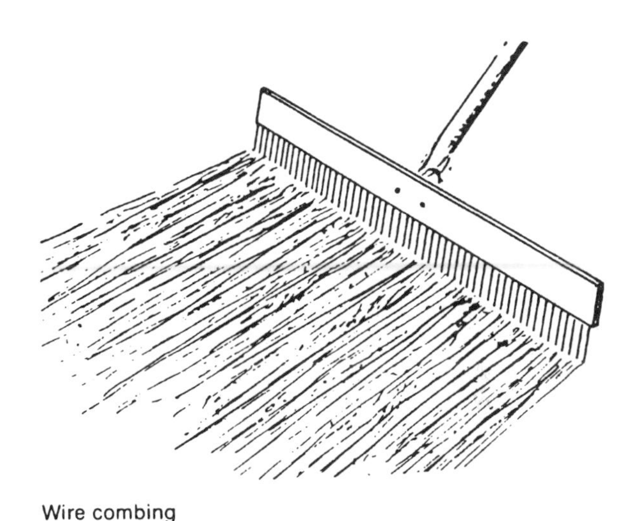

Wire combing

Broom Finish

The broomed texture is very popular. Drag a stiff-bristled broom over the concrete perpendicular to the flow of traffic.

The timing of the finishing process, combined with the bristle stiffness and pressure, determine the depth of the texture.

Coarse

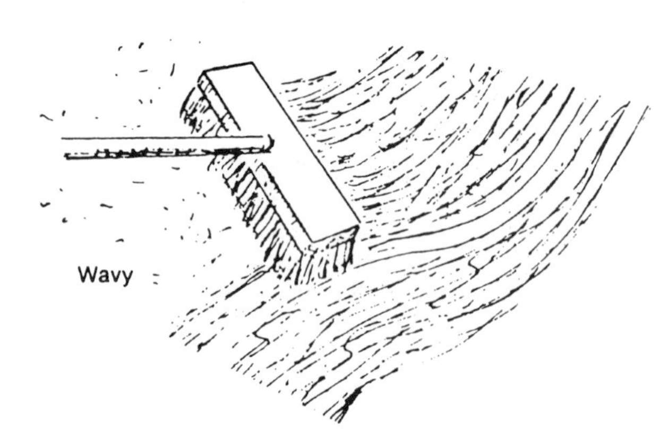

Other Special Finishes

Rock-Salt Texture

Rock salt is scattered over the surface after hand floating or troweling. The salt is rolled or pressed into the surface so that only the tops of the grains are exposed. After the concrete has hardened, the surface is washed and brushed. This process dislodges and dissolves the salt grains and leaves pits or holes in the surface. The grain size could create holes that are about ¼ inch in diameter.

The rock-salt finish is not recommended in areas subject to freezing weather. Water trapped in the recesses of these finishes tends to spall the surface when frozen.

Rock salt texture

Travertine Finish

To attain a travertine finish, apply a dash-coat of mortar over freshly leveled concrete. The dash-coat should be the consistency of thick paint, and usually contains a yellow pigment. Apply it in a splotchy manner with a dash brush so that ridges and depressions are formed. After the coating hardens a bit, flat-trowel to flatten the ridges and spread the mortar. The resulting finish is smooth on the high areas and coarse-grained in the depressed areas. It resembles travertine marble. Many interesting variations of this finish are possible, depending on the amount of dash-coat, color, and troweling.

Travertine finish

Flagstone Pattern

A flagstone pattern is achieved as shown. The joints can be filled with mortar or left open. If they are to be filled, first flood the slab with water to keep it cool and the joints damp. Then remove the water. When the surface is free of water, brush in a bonding grout of portland cement and water mixed to the consistency of thick paint. Next pack in the mortar. A sponge and water are useful for cleaning the joint edges. Ideally, two finishers should work together, one painting in the bonding grout ahead of the mortar and cleaning up the joint edges afterward, while the other finisher concentrates on packing in the mortar firmly and neatly.

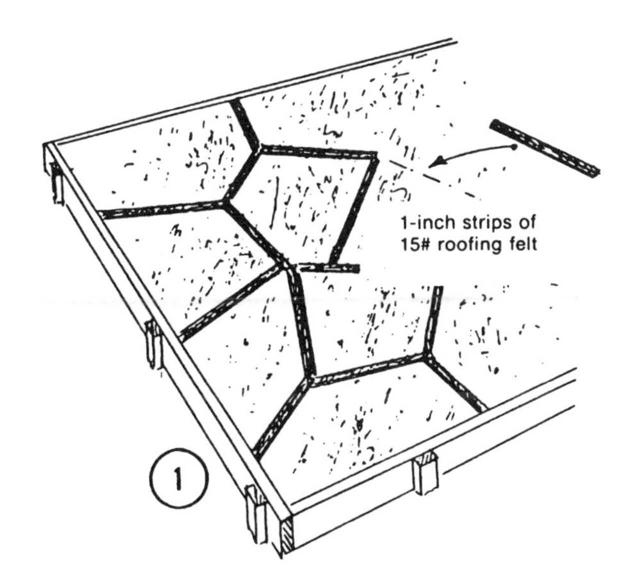

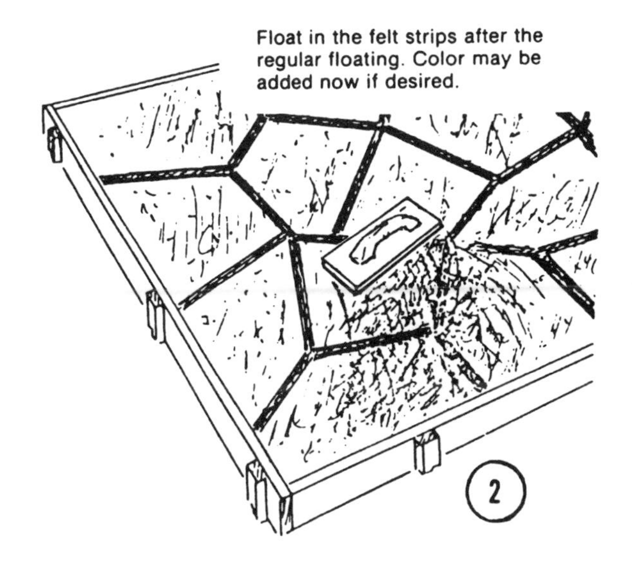

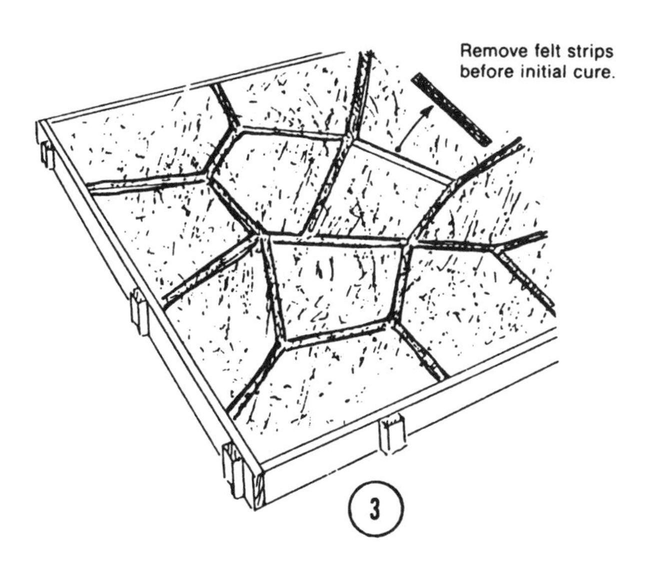

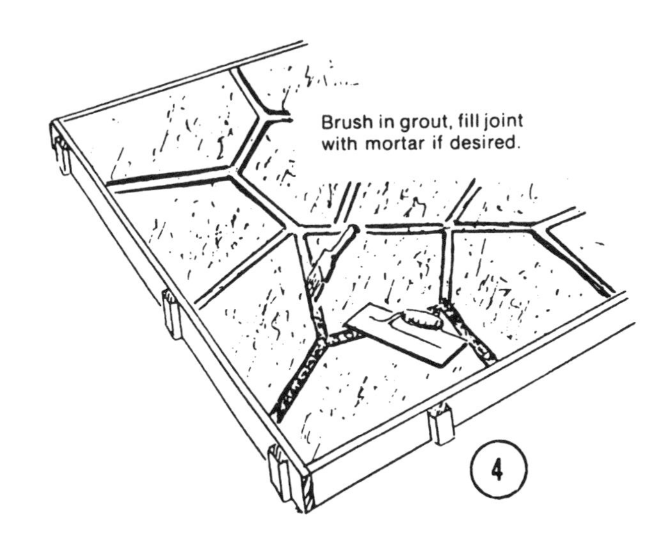

Flagstone Pattern—Alternate Method

Another way to produce a flagstone pattern is with a piece of ½-inch or ¾-inch copper pipe bent into a flat S-shape. After the concrete has been struck off and bullfloated or darbied and no excess moisture remains on the surface, the slab is scored in the desired pattern. This tooling must be done while the concrete is still quite plastic because coarse aggregate must be pushed aside as the tool is pressed into the surface.

The first tooling will leave burred edges. After the water sheen has disappeared, the entire area should be floated and the scoring tool run again to smooth the joints. Floating produces a texture that has good skid resistance and is relatively even, but not smooth. This texture is often used as a final finish.

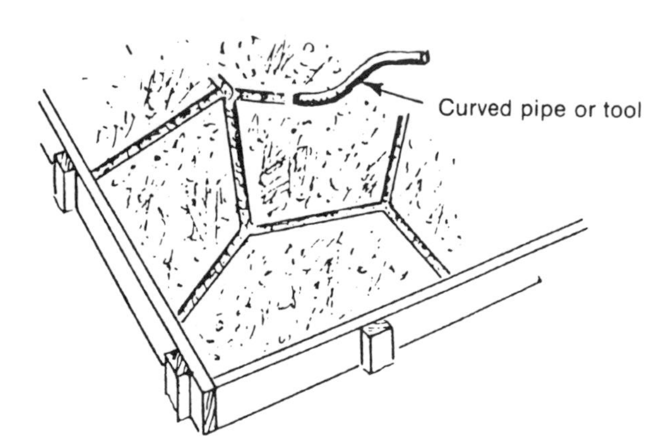

Other Designs

Almost anything may be used to make designs. In addition to the methods shown, pattern stamping can create brick, cobblestone, flagstone, tile, and many other patterns on concrete. Templates for pattern-stamping may be made using a variety of materials, even paper and plastic.

Position each piece of template on the bullfloated slab adjacent to the previous piece until the entire slab is covered with the template.

Embed the paper template using a bullfloat.

Inlaid brick finish after removal of the template and cleaning to remove loose particles from the surface.

Summary for Quality Residential Concrete

The goal of this book is to help builders obtain quality concrete for the many uses in residential construction. Quality concrete results from using good materials, good workmanship, and good planning. Numerous people are needed to supply the materials and perform the labor. Good planning requires that all of these people understand the level of quality expected for each part of the work. This understanding comes from good communications. The residential builder and superintendent must have or must develop good communications skills.

One of the first and most important ways to set the stage for good communications is to call a pre-construction conference. This conference is the time to communicate to each worker and subcontractor the builder's quality requirements. Details, schedules, quality standards, and workmanship are discussed at this meeting. Quality concrete can only be achieved if each person understands that each work activity is important to reach the goal of a quality building.

Glossary of Basic Concrete Terminology

The following general definitions are abbreviated from document ACI 332R-84, published by the American Concrete Institute, and are reprinted with permission.

Admixture - A material other than water, aggregates, and hydraulic cement used as an ingredient in concrete. (See also **Air entrainment** and **Water-reducing admixtures**.)

Aggregate - The natural or crushed stone and sand which are the ingredients that make up the largest fraction of most concrete mixtures.

Air entrainment - The intentional incorporation of minute air bubbles in concrete to improve durability to freezing and thawing exposures or to improve workability. Accomplished either by use of an air-entraining admixture or an air-entraining cement.

Anchor - A steel unit set in concrete (sometimes by attaching to the formwork) for later use in attaching something else to the concrete.

Backfill - The soil that is compacted into place to correct for over-excavation.

Beam - A structural member subject to bending, generally used horizontally to support a slab or wall.

Bleeding - The movement of water within fresh concrete toward the top surface and its collection there, caused by settling of the solid materials.

Blistering - Formation of thin, raised flaws on the surface of concrete during the finishing operation or soon afterward, which are not always easily noticeable before the concrete hardens.

Blowup - The rise of two concrete slabs where they meet as a result of greater expansion than the joint between them will accommodate. Blowups are likely to occur only in unusually hot weather at locations where joints have become filled with incompressible materials. They often result in cracks on both sides of the joint and parallel to it.

Bugholes - See **Surface air voids.**

Bulkhead - A partition inserted in formwork to block fresh concrete from flowing into another section of the formwork.

Bullfloat - A T-shaped tool with a large flat blade attached with a hinged joint to a long handle used to smooth the surface of freshly screeded concrete flatwork.

Calcium chloride - A salt, sometimes supplied in solution or found as an ingredient in admixtures, used to accelerate the setting or strength gain of concrete; may contribute to rusting of reinforcing steel or (only when used as an admixture) discoloration of flatwork surfaces.

Cast-in-place - Concrete deposited in the place where it is required to harden rather than precast and moved into position after curing and hardening.

Caulking - The filling of a joint with a material suitable for sealing out dirt and moisture. Better-grade materials used for this purpose are commonly known as sealants.

Cement - The powder (usually portland cement) which, when mixed with water and aggregate, slowly reacts chemically with the water to form the bonding agent that holds the aggregate together, producing concrete. (The term "cement" should not be misused to refer to concrete.)

Chair - A device used to hold reinforcing bars in their proper position during placing and working of concrete.

Chute - A sloping trough down which concrete moves from the ready mixed concrete truck to a receptacle or form.

Clay - A very fine natural soil with plastic properties when moist. Some soils contain clay mixed with other ingredients.

Cold joint - A joint, usually visible, in a concrete wall or floor where the fresh concrete has bonded imperfectly or not at all to the previously placed concrete because too much time has elapsed between placements.

Column - A concrete member (usually vertical to support a floor or roof) with slender proportions that takes compression loads.

Concrete - A composite material made of portland cement or other hydraulic cement, aggregate, water, and sometimes admixtures, which hardens when the cement reacts chemically with water.

Construction joint - The plane where two successive placements of concrete meet but do not bond cementitiously. Usually it is only necessary to use a keyway for load transfer across the joint, but sometimes dowels or reinforcing steel are required to cross the joint to hold the concrete on both sides together.

Control joint - A joint purposely designed to accommodate movements in concrete inevitably caused by temperature changes and drying shrinkage. Made by forming, tooling, or sawing a groove in a concrete structure, this creates a weakened plane so that cracking will occur along this predetermined line and not at random locations. Also called a **contraction joint**.

Cure - To retain moisture in concrete for a prescribed period and at a desirable temperature to allow the cement to chemically react with water and reach the required strength level and other desirable properties of concrete.

Curing compound - A liquid that can be applied to the surface of newly placed concrete to retain water in the concrete long enough for it to be cured.

Curling - The turning up of the edges and particularly the corners of a slab caused by the drying or cooling of the top surface faster than the bottom surface.

Darby - A long, straight, flat surface with inclined handle used in the early stage of leveling operations on concrete slabs.

Deformed bar - A steel reinforcing bar with raised deformations on the surface to provide an interlock with the surrounding concrete.

Dowel - A steel pin or bar extending into two adjoining portions of a concrete construction to connect them and transfer load.

Durability - The ability of concrete to resist weathering action, chemical attack, abrasion, and other conditions of service.

Dusting - The appearance of a powdery material on the surface of hardened concrete coming from the concrete itself.

Edging - The operation of tooling the edges of a fresh concrete slab to provide a rounded corner.

Efflorescence - A deposit of salt or salts, usually white, formed on a surface. The substance is one that has emerged in solution from within the concrete and has been deposited by evaporation.

Expansive soil - A soil subject to considerable increase in volume change with resulting uplift or distortion of concrete members. This is a severe problem in a few areas of the United States.

Fault - A vertical movement of a slab or other member adjacent to a joint or crack so that there is an abrupt change in surface elevation from one side of the joint to the other.

Fill - See **Backfill**.

Finishing - Operations such as floating and troweling that produce a surface of the desired smoothness, density, and flatness; these operations are made easier by a well-proportioned mix that is adequately cohesive and plastic.

Flatwork - A general term that encompasses floors, patios, walks, driveways, and other slabs-on-ground.

Floating - The operation of finishing a fresh concrete slab surface by using a hand or power float.

Flow line - A detectable line on a concrete wall or column usually departing somewhat from horizontal that shows where the concrete in one placement has flowed horizontally before the succeeding placement has been made. Good concreting practices should eliminate most evidence of flow lines.

Flowing concrete - Concrete to which has been added water-reducing, set-controlling admixture or admixtures to produce a temporarily high slump to aid in placing and consolidation.

Fly ash - A finely divided glass-like powder recovered from the flue of a coal-burning industrial furnace. It is sometimes used as a mineral admixture in concrete to react with the cement and modify or enhance the properties of the concrete.

Footing - The part of the foundation that spreads and transmits the load to the soil.

Form - A large mold of lumber or prefabricated elements set up to support and contain concrete until it has gained sufficient strength to be self-supporting.

Form coating - A liquid that may be applied to the surface of the form for one or more of the following purposes: to protect the form surface and give it long life; to retard the set of the surface of the concrete to make it easy to expose the aggregate at a later time; or to promote the ease with which formwork can be removed (stripped) from the concrete. (Form coatings for the latter purpose are also called **form release agents**.)

Form sealer - A liquid applied to the surface of a form to reduce or overcome its absorptivity of moisture from the concrete.

Form spacer - A temporary wood or steel insert placed between side panels of a form to resist the tension of the ties until concrete has been placed. (Also called a **form spreader**.)

Form tie - A manufactured steel wire, bar, or rod specially designed to prevent concrete forms from spreading due to the fluid pressure of freshly placed concrete.

Grade - The prepared surface on which a concrete slab is cast. To prepare a plane surface of granular material or soil on which to cast a concrete slab.

Girder - A large beam, usually horizontal, that serves as a main structural member.

Grade tamper - A hand tool or powered device for compacting the grade by a pounding action.

Grout - A mixture of cement and water and sometimes fine sand proportioned to produce a pourable consistency without separating.

Honeycomb - Voids left in concrete where cement and sand particles have not filled the spaces among the coarse aggregate particles.

Insert - Anything other than reinforcing steel that is rigidly positioned within a concrete form for permanent embedment in the hardened concrete.

Isolation joint - A built-in separation between adjoining similar or dissimilar elements of a concrete structure, usually a vertical plane. It can also be used to separate two concrete structures such as a walk and a driveway or a patio and a wall. Its purpose is to prevent movements of the individual parts from causing cracks in the concrete. (Also called **expansion joint**.)

Jointing - The process of producing joints in a concrete slab with a metal hand-tool made for the purpose.

Keyway - A recess or groove made in one placement of concrete that is later filled with concrete of the new placement so that the two lock together.

Lateral pressure - Pressure exerted in a horizontal direction against formwork by the hydraulic fluid pressure of fresh concrete.

Lintel - A horizontal structural element above a window or door to support the wall above.

Load-bearing wall - A wall designed and built to carry vertical and shear loads in addition to its own weight.

Mobile placer - A small belt conveyor mounted on wheels that can be readily moved to the jobsite for conveying concrete from the ready mixed concrete truck to the forms or slab.

Monolithic concrete - A large block of cast-in-place concrete containing no joints other than construction joints.

Muriatic acid - A mineral acid more properly known as hydrochloric acid, available at most hardware stores, sometimes used for cleaning or acid-etching concrete or removing efflorescence.

Penetration - An opening through which pipe, conduit, or other material passes through a wall or floor.

Permeable concrete - Concrete with higher-than-normal susceptibility to having water pass through it. The permeability of high-quality concrete may be only one-millionth that of low-quality concrete.

Placeability - Fresh concrete's capability of being easily placed and consolidated, largely dependent on composition and proportions. Concrete that has good placeability is likely to have good finishing qualities, though these two qualities are not identical.

Plastic shrinkage - The shortening of the surface of fresh concrete from rapid evaporation of moisture due to low humidity, high winds, high temperature, or a combination that often leads to the creation of cracks before the concrete has been finished.

Ponding - The creation and maintenance of a pond of water on the surface of a concrete slab for the purpose of curing.

Popout - A small fragment of concrete that has broken away from the concrete surface because of internal pressure, leaving a conical pit.

Portland cement - A hydraulic cement conforming in composition and properties to the requirements of ASTM Standard C150. There are many different brands of portland cement, all of which conform to the specification.

Pozzolan - A finely divided material which is not itself a cement but which reacts chemically with the products of hydration of portland cement to form a cementitious binder. Sometimes used as an admixture in concrete to modify or enhance the concrete properties.

Prepackaged concrete - Bagged material consisting of a dry pre-proportioned mixture of cement, coarse and fine aggregate, and sometimes admixtures, usually used for small jobs. Water is added at the mixer.

psi - Abbreviation for pounds per square inch. (In the new SI metric system, units of pressure are expressed in megapascals.)

Pump - A specially designed machine capable of forcing fresh concrete through a pipeline or hose having a diameter of about 3 to 6 inches.

Ready mixed concrete - Concrete batched in a concrete plant and mixed in a plant or in the truck mixer that delivers the concrete in a plastic, unhardened state.

Reinforcing bars - Steel bars embedded in concrete to act with the concrete in resisting forces. (See also **Deformed bars**.)

R-value - A standard measure of the resistance that a material offers to the flow of heat.

Sand streaking - A line of exposed fine aggregate on the surface of formed concrete caused by bleeding of water from the concrete.

Scaffolding - A temporary structure to support a platform for workers, tools, materials, and carts; or to support formwork for an elevated slab or beam of concrete.

Scaling - Flaking or peeling away of a surface portion of hardened concrete.

Screed - A firmly established grade strip or side form that, in combination with another strip on the other side, serves as a guide for striking off the surface of a concrete slab to the desired level.

Sealant - Extensible material used to seal a joint to exclude water and solid foreign materials.

Sealer - A liquid composition applied to the concrete surface to diminish the absorption of water, solutions of deicers, or other liquids.

Segregation - Partial separation of the various materials that make up concrete during the transporting, handling, and/or placing operations, resulting in a non-uniform product.

Seismic zone - An area of the country in which earthquake intensity is likely to fall within the designated range for that zone as specified in standard building codes.

Shale - A laminated sedimentary rock that is not very hard and can be readily reduced to clay and silt.

Sheathing - The material used to form the contact face of forms.

Shoring - Props or posts of timber or other materials used temporarily to support concrete framework.

Silt - A granular material formed from rock disintegration small enough to pass through a sieve with 200 openings to the inch.

Slab - A flat or nearly flat horizontal surface of plain or reinforced concrete used as a floor, roof, pavement, patio, or walk.

Sleeve - A pipe or tube passing through formwork for a wall or slab through which pipe, wires, or conduit can be passed after the forms have been stripped.

Slump - A simple, convenient measure by a standard test method of the consistency of freshly mixed concrete.

Spalling - The breaking away of a small shape or chunk of concrete, usually by expansion from within the larger mass.

Stress - The intensity of the internal force within concrete. The stresses usually considered are those of tension or compression, although stresses of torsion or shear can be important. Stress is expressed mathematically in terms of force per unit area.

Striking off - The process of shaping the surface of a freshly placed concrete slab by using a straightedge tool or a special machine to level it to the elevation of the screeds.

Stripping - The process of removing forms from concrete after it has hardened.

Superplasticizer - See **Water-reducing, set controlling admixture**.

Surface air voids - Small round or irregular cavities usually not more than 5/8 inch in diameter resulting from air bubbles trapped in the surface of formed concrete during placement and compaction. Sometimes called bugholes.

Tie-bar - A deformed bar embedded in concrete at a joint to hold the abutting edges together.

Tolerances - The variation permitted from the dimension given, for example, from the planeness of a floor or from the location or alignment of a concrete wall.

Troweling - Smoothing and compacting the surface of a concrete slab by strokes of a trowel.

Unbalanced fill - The height of outside finish grade above the basement floor or inside grade.

Vibrator - An oscillating power tool used to agitate fresh concrete to eliminate entrapped air (but not entrained air) and bring the concrete into intimate contact with formed surfaces and embedded materials.

Wale (or **Waler)** - A long horizontal formwork member used to hold vertical framing members in place.

Water-reducing, set-controlling admixture - Any of a number of chemical materials or combinations of chemical materials added to concrete to enhance the performance of concrete in both the plastic and hardened states. ASTM C 494 outlines "normal range" materials (Types A through E) and "high-range" or "superplasticizing" materials (Types F and G).

Waterstop - A thin sheet of metal, rubber, plastic, or other material inserted in a form across a joint to obstruct the flow of water through the joint.

Welded wire fabric - A mesh made of longitudinal and transverse wires crossing at right angles and welded together for use as reinforcement in concrete. Supplied in either sheets or rolls.

Workability - The ease of response of concrete in mixing, placing, compacting, and finishing.

Sources of Additional Information

Builders seeking information about specific aspects of building with concrete often obtain assistance through contacts with their local home builders associations and their concrete suppliers. Additional information and publications are available from professional organizations, including—

National Association of Home Builders
Technology and Codes Department
1201 15th Street, NW
Washington, DC 20005-2800
(800) 368-5242

NAHB Research Center
400 Prince Georges Boulevard
Upper Marlboro, MD 20772-8731
(800) 638-8556

Problem Clinic Editor
Concrete Construction
The Aberdeen group
426 S. Westgate Street
Addison, IL 60101
(800) 323-3550

American Concrete Institute
Member/Customer Services
P.O. Box 19150
Detroit, MI 48219
(313) 532-2600

AFM Corporation
(for information on polystyrene forms)
P.O. Box 246
Excelsior, MN 55331
(800) 255-0176

National Ready Mixed Concrete Association
Engineering Department
900 Spring Street
Silver Spring, MD 20910
(301) 587-1400

Editor, *Concrete Technology Today*
Portland Cement Association
5420 Old Orchard Road
Skokie, IL 60077
(708) 966-6200

Wire Reinforcement Institute
1101 Connecticut Ave., NW
Washington, DC 20036-4303
(202) 429-5125

Numerous books, periodicals, and videotapes are available on specific aspects of concrete construction. The following items provide a sampling of helpful additional resources—

ACI 214.4R, "Guide for the Use of High-Range Water-Reducing Admixtures (Superplasticizers) in Concrete," *Concrete International,* April 1993.

ACI, *Concrete Craftsman Series: Slabs on Grade,* American Concrete Institute, Detroit, MI, 1982.

ACI 332R-84, "Guide to Residential Cast-in-Place Concrete Construction," Committee 332, American Concrete Institute, Detroit, MI, 1984.

ACI videotape CP-5, "Field Tests for Quality Control of Fresh Concrete," American Concrete Institute, Detroit, MI, 1987.

Jackson, W.P., *Building Layout,* Craftsman Book Company, 1990.

PCA EB001.13T, *Design and Control of Concrete Mixtures,* 13th Edition, Portland Cement Association, Skokie, IL, 1988.

PCA PA122.05H, *Cement Mason's Manual,* Portland Cement Association, Skokie, IL, 1990.

PCA PA124H, *Finishing Concrete Slabs with Color and Texture,* Portland Cement Association, Skokie, IL, 1991.

PCA SP038H, *The Homeowner's Guide to Building with Concrete, Brick, and Stone,* Portland Cement Association, Skokie, IL, 1988.

NRMCA, "Solid Concrete Basement Walls," National Ready Mixed Concrete Association, Silver Spring, MD, 1977.

More Construction Books From Home Builder Press

Beam Series
In these five volumes you'll find a direct way to determine the most appropriate and economical fabricated structural beam for specific uses. Covers wood, plywood, plywood box, steel wood I-beams, and flitch plate and steel I-beams.

Building with Alternatives to Lumber and Plywood
Put together by the NAHB Research Center, this book gives you needed information about currently available or emerging alternatives to lumber and plywood, including engineered wood products, laminated fiberboard structural sheathing, steel framing, foam core structural sandwich panels, concrete wall and floor systems, concrete block, and more. Basic descriptions, applications, advantages and disadvantages, and some cost information included.

Cost-Effective Home Building
The NAHB Research Center has revised and expanded this book (formerly *Reducing Home Building Costs with OVE Design and Construction)* to bring you up-to-date information and techniques on how to reduce costs without compromising quality. Learn how to use construction materials and resources more efficiently from the planning stage through finishing.

Energy-Smart Building for Increased Quality, Comfort, and Sales
Builder Philip Russell presents what you need to know to build—and sell—energy-efficient homes, from the thermal envelope to marketing tips and sources for cutting-edge materials and technologies.

Land Development
This comprehensive book covers everything you need to know about developing land, including site selection and analysis, local land use regulations and plan processing, selection of housing types, earthwork and stormwater management, water and sewer, residential streets, energy conservation, landscaping, and information on market research and financing.

Scheduling for Builders and *Estimating for Builders*
Two of our best selling books, *Scheduling* and *Estimating* provide practical, reliable guidance to managing the thorny details of costing jobs and juggling schedules. These two books will help you get the work done on time and within budget.

Production Checklist
This comprehensive checklist (with related schedules) organize day-by-day, step-by-step construction procedures to assist you as you confirm start dates, monitor progress, predict completion dates, meet your production schedules, and produce a quality finished product—every time.

Wood Frame House Construction
With this book you'll build smart and improve the quality of your wood frame homes from conception to completed structure. Special features such as flooring and maintenance are included along with the basics. Fully illustrated.

To order any of these books, or request an up-to-date catalog of Home Builder Press titles, write or call:

Home Builder Bookstore
1201 15th Street, NW
Washington, DC 20005
(800) 223-2665

NAHB Members Receive a 20% Discount on All Books